Gender, Class and Power

Tricia Dawson

Gender, Class and Power

An Analysis of Pay Inequalities in the Workplace

palgrave
macmillan

Tricia Dawson
Keele Management School
Keele University
Keele, UK

ISBN 978-1-137-58593-6 ISBN 978-1-137-58594-3 (eBook)
https://doi.org/10.1057/978-1-137-58594-3

Library of Congress Control Number: 2018936137

Printed on acid-free paper

This Palgrave Macmillan imprint is published by the registered company Macmillan Publishers Ltd. part of Springer Nature.
The registered company address is: The Campus, 4 Crinan Street, London, N1 9XW, United Kingdom

Acknowledgements

Special thanks to John Abbott; Professor Carole Thornley; the former Graphical, Paper and Media Union (currently a sector of Unite); and, in particular, all those who gave their time to provide me with the information that I have used in this book. Any mistakes remain my own.

Contents

Abbreviations

ASLP	Amalgamated Society of Lithographic Printers
ATCA	Art, Technical, Clerical and Administrative (trade group board of the NGA)
BDC	Biennial Delegate Conference; pre-1991, the SOGAT conference, 1991 onwards the GPMU conference
BDM	Biennial Delegate Meeting; the NGA conference
BFMP	British Federation of Master Printers
BPIF	British Printing Industries Federation
EC	Executive council
EHRC	Equality and Human Rights Commission
EOC	Equal Opportunities Commission
EPA	Equality policy adviser
EU	European Union
F/MoC	Father or mother of the chapel
GFC	Global financial crash
GPMU	Graphical Paper and Media Union
GPR	Guaranteed proportional representation
HMSO	Her Majesty's Stationery Office
LTS	London Typographical Society
MEG	Minimum earnings guarantee
NATSOPA	National Society of Operative Printers and Media Personnel (also NSOP&A)
NEC	National Executive Committee

NEDO	National Economic Development Office
NGA	National Graphical Association
NPA	Newspaper Proprietors' Association
NS	Newspaper Society
NUJ	National Union of Journalists
NUPBPW	National Union of Printing, Bookbinding and Paper Workers
NUWDAT	National Union of Wallcoverings, Decorative and Allied Trades
P&KTF	Printing and Kindred Trades Federation
PPITB	Printing and Publishing Industries Training Board
SLADE	Society of Lithographic Artists, Designers, Engravers and Process Workers
SOGAT	Society of Graphical and Allied Trades
SPEF	Scottish Printing Employers Federation
TA	Typographical Association
TUC	Trades Union Congress

List of Figures

List of Tables

1

Introduction

Forty years after the implementation of the Equal Pay Act, gender pay discrimination continues unabated. At the commencement of this research, the researcher was equality policy adviser (EPA) to the Graphical, Paper and Media Union (GPMU), with responsibility for equalities nationally. This research was driven by the need to understand the totality of barriers to achieving equal pay in the British printing industry. The union had worked hard on many policy areas that contribute to challenging the pay gap—maternity, paternity, parental leave and pay; harassment and bullying (under the banner of Dignity at Work); childcare; and equal opportunities. Training courses had been run with women members, union local negotiators and male workplace representatives to try to increase gender awareness. Greater publicity was given to these issues in union communications, at conference fringe meetings and in conference debates. Radical structural change ensured women were included at national level in all decision-making bodies of the union. None of this had made a marked difference to women's access to jobs or their access to equal pay. It was time to tackle equal pay directly. In a union that had an 83 per cent male membership, it was hard to see how to build momentum

T. Dawson, *Gender, Class and Power*, https://doi.org/10.1057/978-1-137-58594-3_1

for this crucial policy area or how solidarity could be called for when it seemed that few thought there was an equal pay issue; it was more about access to Class 1 jobs, not about pay.

It was in these circumstances that it seemed necessary to investigate the totality of barriers that affected women's access to pay and jobs—maybe this would throw light on points of leverage that could be developed into a meaningful campaign that might begin to shift barriers that had existed since the printing industry's inception in the fifteenth century. This research was designed to examine the persistence of these barriers. It commenced in 2000 and the first phase was completed in 2007, during which time the GPMU ceased to exist as an independent union, merging with Amicus in 2004, then Unite in 2007. Subsequently, the author, as a professional academic, updated the research to take account of events leading up to, and after, the great financial crash (GFC) of 2008. While the GPMU supported and helped fund the original research, the second phase was unsupported and conducted as time allowed between 2014 and 2017. This second phase revisited some of the union officials who were originally interviewed but also included women activists, who were excluded in the original research for lack of time. This provided an opportunity to fill a gap in the research and to allow for reflection on the GPMU's approach to equalities that has afforded a useful corrective to the original research and a more balanced account. Time has also allowed the researcher to reflect on and refine the original approach and understanding of the barriers identified.

During the two decades before the original research, manufacturing industry had altered considerably. Employment had greatly diminished, shifting to the service sector, trade unions were contained and industrial action diminished drastically. In printing, there was also significant technological change affecting the labour process. Despite these changes, there does not appear to have been any corresponding change in the position of women in these jobs, even by 2017. The aims and objectives of this research were, thus, to determine the causes of the gender pay gap in the printing industry, why it is so resistant to change and under what conditions change may be effected.

While the ultimate aim of this research is to uncover the unequal rewards women receive from paid employment, there is also an appreciation that

> [m]oney is only one form of gender power relations in the workplace and in the labour market … [there is also, for instance] equality of opportunity to pursue careers, or to participate fully in trade unions. (Wajcman 2000: 187)

As such, this book considers the historical development of women's work in the printing industry, followed by an analysis of work organisation, employment policies and trade union democracy in contributing to unequal pay. This culminates in a discussion of the pay structure and the contribution of pay determination processes in embedding unequal pay outcomes. It should be noted, however, that while a national agreement controlled the pay structure at the time of the original research, this agreement foundered in 2010 and no longer exists. Employers are free to impose pay in many companies, although where chapel (workplace branch) organisation remains strong management prerogative may still be contained.

The researcher's own experience and the focus of earlier research affected the choice of sector. In particular Cockburn had published a highly regarded study of newspaper compositors and their response to radical technological change in Cockburn (1983). This account dealt mainly with men because it focused on a sector from which women were, traditionally, excluded: newspaper production. It provided a sensitive and detailed analysis of men's relationship to new technology and to women, highlighting the importance of masculine ideology. However, it remains the case that women's responses were only marginally important in that research. Within the print unions, the researcher recalls hostility after the publication of Cockburn's book. In the second phase of research, the researcher questioned a woman official who had been around at that time, someone who would identify as a socialist feminist (R2[4]). She, too, felt some hostility because, it seems, union officials felt the study did not do enough justice to class relations in the industry. This research attempts to avoid that pitfall.

Conversely, this research is focused on the sector of the industry where women were most often found—the general printing industry. This sector covers 'packaging, advertising materials, security printing, business forms, books, periodicals, magazines, stationery and catalogues' (Gennard and Bain 1995: 22–23). The labour process is still recognisable from 200 years ago, comprising a basic division between pre-press (production of material for printing), printing (the actual process of printing onto various materials) and the bindery (all activities required to produce a finished product). Despite technological change, print production still maintains this basic division, which is highly gendered. Women had entered pre-press areas at various points (with limited success, see Chap. 5), whereas they were excluded before changes in technology made their typing skills useful for employers. Re-segmentation has been the issue here, maintaining degrees of gender segregation that affect pay. Women continue to be found in the largest numbers in binderies, although their jobs have greatly reduced in number. Here they are segregated into jobs regarded as unskilled and commanding the lowest pay rates. Meanwhile, in the press rooms where the printing is carried out, women have barely made their presence felt, even today.

Work organisation processes are crucially linked to bargaining processes in this sector. Multi-employer bargaining had dominated general printing, in contrast to many other sectors (Roe 2003). The national agreement for the sector had considerable importance, not just for those directly party to it but also non-union and non-federated companies. This meant that even where the union was absent or had minimal membership, or where companies were not members of the employers' association, the British Printing Industries Federation (BPIF), they often benchmarked wages and conditions against the agreement. This was important because the national agreement significantly influenced the allocation of people to both jobs and pay grades (Craig et al. 1984). Women's pay was institutionalised by such arrangements. This is no longer the case, and all workers' wages and conditions were adversely affected, especially following the GFC. Traditionally, there was also an expectation of second-tier bargaining, which was waning even at the time of the original research. But there is some recent evidence that strong chapels are beginning to recoup some of their lost benefits. These chapels are likely to be entirely male, so the pay gap could begin to widen again. This book attempted to tease out these processes to identify their gendered effects.

This account also needs to be seen in the context of the decline of traditional printing as a major industrial sector in Britain. Economic restructuring and, in particular, increased international competition were affecting the industry, even during the original research, but the subsequent economic upheaval exacerbated these trends. Technological change reduced the need for labour and facilitated the movement of production overseas. Much of the actual printing process is undertaken in other countries, while the remaining work has moved increasingly into the communications[1] sector. This has greatly reduced unionisation. Whereas there were some 280,000 members claimed at the time of the merger to form the GPMU, there are now only around 50,000—and this includes IT workers who are not in traditional print areas at all. Women's employment suffered relative to that of men, reducing their numbers even further. This has made the problem of challenging gendered processes even harder.

Previously technological change had tended to result in the unions 'following the job' to the new equipment. For instance, the decline of letterpress printing led the unions to focus attention on lithographic printing, the new technology of the period. Similarly the decline of compositors' jobs led to a focus on typesetting and attempts at retaining it as a male occupation. The only times this has not been evident have been when this has involved movement into a feminised sector, such as the operation of office-based printing equipment or, latterly, the move into publishing and information technology. The decline of the GPMU can be, at least partially, attributed to this strategic failure—although there were, and still are, attempts being made to gain organisation in what was termed 'new media'.

This book attempts to provide as comprehensive a picture as possible of both the economic and social processes that produce gendered outcomes. Specifically, it considers the unequal allocation of power resources that generate and sustain women's invisibility in the workplace, in the trade union and in bargaining processes, rendering them seemingly irrelevant in all three areas. This theory rests on the elements of power identified by Lukes (2005), especially hidden forms of power that tend to 'naturalise' the status quo. The vehicle for this exercise of power is a set of power resources, providing unequal access to power, dependent on gender. This examination of power resources rests on the idea of class and gender operating as dynamics, rather than fixed structures, to embrace the notions of 'continuity within change, order within variability, fixity within fluidity' (Bradley 1996: 7). Bradley's (1999) multi-dimensional model of power resources is utilised to investigate the unequal use of power to sustain the status quo. It is extended to include legal power, a resource accessed through women's campaigning. The model also attempts to analyse how access to certain key power resources extends access to further power resources that subsequently operate to reinforce each other and embed gendered power relations. Processes of reproduction, rationalisation and resistance in the workplace and, to some extent, the home (Collinson et al. 1990) are analysed. The result is to demonstrate how unequal access to power resources, including temporal differences in such access, results in relative privilege for both employers and male workers that are so embedded they appear natural and, therefore, invisible. Consequently, women struggle to challenge existing workplace relations, especially given their late entry into paid work. To the extent women have successfully achieved legal protection, equal opportunities policy initiatives and structural change in unions, it is argued here that gendered power relations give rise to a process of 'proceduralisation' that averts transformational change.

Intersectionality can be understood as the identification of, for example, race, class, gender, age, sexual orientation and religion as interlocking elements of 'one overarching structure of domination' (Collins 1990: 222). This book embraces the idea that 'gender cannot be understood in isolation from other social constructions such as 'race' and class' (Holgate et al. 2006). Throughout the following chapters, both gender and class relations are given their due weight in the analysis. The result is a new synthesis of existing theoretical strands to expose both the visible and invisible processes of power relations in sustaining women's (relative) invisibility and resulting, apparent, irrelevance. The tenacity and intractability of these processes are revealed throughout the following chapters.

Research Process

This section briefly identifies the philosophy behind the analysis contained in this book and the methods used. While this research was undertaken from an industrial relations (IR) perspective, the aim was to ensure women occupied a central position in it to avoid the gender blindness that is traditional in IR research (Ledwith and Hansen 2013). IR research is interdisciplinary and draws on many cognate disciplines to gain an understanding of workplace relations. In this research economics, sociology, history and law all play a part in exposing gender relations in printing. Intersectionality is a key aspect of this approach—an attempt to do justice to both class relations and gender relations and how they interrelate. While inevitably such research identifies tensions in gender relations, there was no desire to demonise men, but to understand the pressures on both women and men in the workplace and the potential to overcome them. Employers play a key role in provoking and stoking these tensions, and it was crucially important to expose these processes.

A key issue highlighted in this book is the maintenance of invisibility. This does not just relate to the invisibility of women and their issues but the invisibility of employers' mechanisms for maintaining gender relations and the invisibility of relative male privilege among workers. Male identity is taken as the norm so often that it is not seen as an identity at all. As such all other identities are measured against it and, increasingly, criticised for disappearing into a post-structural morass (see, e.g., Fraser 2008). Such an argument evokes the suspicion of gender analysis that tended to envelope IR research, which was, traditionally, more concerned with class antagonisms (Foster and Williams 2010). But this book, it will be seen, exposes the persistence of discrimination in the industry and how closely this is bound up in male identity—in terms of both employers and male workers.

Ultimately, the aim of the research contained in this book was emancipatory, aimed at transforming and empowering those involved (and beyond) to change their lives (Gottfried 1993). Given the author's role as EPA for the GPMU, it could be argued that this aim also informed the researcher's 'day job', feeding into processes of policy development and implementation along the way and so was not 'just talk' (ibid.). Instead it proceeded from a 'political commitment to produce useful knowledge that … [would] make a difference to women's lives' (Letherby 2003: 4) and as such affected both the 'why' of choosing the research topic and the 'how' of constructing the research (Hesse-Biber and Yaiser 2004).

In this light, the research was explicitly concerned with what was problematic for women, rather than men (Harding 1987). While it is true that men and women share many workplace problems, 'it is indisputable that gender has an important impact on the experience of work' (Holgate et al. 2006: 310), but this is not immediately obvious in IR research. This was largely true of printing industry research, too, with the exception of the studies of unions conducted by Ledwith et al. (1985), Ledwith (1991), Colgan and Ledwith (1996) and Ledwith and Colgan (2002). These studies have provided invaluable material but concentrate primarily on union dynamics, rather than broader industry developments and employer strategies.

However, in looking at research questions from a gender perspective, the researcher can be accused of relativism (Harding 1987), of taking a post-modernist or post-structuralist approach. Post-modernism eschews major theories reducing everything to the local, while post-structuralism relies heavily on discourses to carry the full weight of explanation, ignoring structural issues (Connell 1987: 242). Like Bradley (2007) this study ultimately does not find the individualising tendencies of these approaches helpful and, conversely, while recognising the continuing importance of structures, is keen to include an appreciation of the protagonists' ability to act for themselves.

The research utilised both quantitative and qualitative research methods. The contribution made by quantitative methods in this research was to provide for the widest possible gathering of information across the sector on all the key aspects of employment, for example, recruitment, training, promotion, pay, treatment of workers, domestic responsibilities/work interface and union/member relations. Data on different experiences of work, by sex, would be available for the first time for this sector on a broad basis. Qualitative methods encompassed both extensive documentary research and interviews with specific groups: employers, union officials and union members.

Documentary evidence was a crucial element of the research strategy, much of it preceding and informing the development of the survey and interview schedules. Such evidence requires careful handling to identify the potential biases in documents, especially for areas of history that have been largely ignored. For example, existing histories may reveal as much from what is left out, or how it is reported, as from what is actually presented. Such was the case with some of the histories used here (see, e.g., Roe 1996). Documents may also operate as 'guides for action … [and where they are] the outcome of collective decision making through negotiation or consultation, they become important tools in the practice of industrial relations' (Healy 1997: 209), as is recounted in Chap. 7 with the implementation of the national agreement.

The contribution made by the interviews was vital in developing theory, going deeper into the barriers to eliminating discrimination and also allowing assessment of how interviewees' backgrounds might contribute to their attitudes and behaviours. In particular, this provided a rich source of evidence for men's and employers' resistance to change despite the success of women's campaigning. Crucial issues of process, such as collective bargaining and work organisation, were teased out. In particular, it was the main method of giving women a voice in identifying their own problems and of relating these to the attitudes exhibited by men in the workplace. One aspect of the terminology used in explaining some aspects of the research needs explaining. It has become more common to refer to the relevant union officials as paid officials rather than full-time officials (FTOs), in order to recognise that the full-time model of union office may disadvantage women and there is no reason to insist on full-time hours. However, it is used knowingly in this book to attest to the fact that such officials were, in fact, full-time.

Given the author's role as EPA, participant observation was also an important aspect of the research. While this requires strenuous efforts on the part of a researcher in maintaining 'a high degree of self-awareness (Bryman and Bell 2011: 440) to counteract the effects of such familiarity with the research subjects, it also allowed the author to utilise her experience to 'inject some judgement about employer and Federation claims … [and therefore] the gap between rhetoric and reality' (Roe 2003: 10–11).

The purpose of this variety of methods was to examine practice and process in gender power relations at work. It must be acknowledged, however, that capturing practices which tend to occur 'unreflexively … [that] happen fast, are "in action" and occur on many levels' (Yancy Martin 2003: 344) is very difficult and also hard to express clearly. But as Joan Acker (1990) has argued, making such practices visible is essential to challenging discrimination. Creating visibility of practice is the essence of this study and has powerful effects, as can be seen in the news every day at the time of writing, with the erupting sexual harassment scandals emanating from the entertainment and political arenas (see, e.g., bbc.co.uk and other media outlets October/November 2017).

Organisation of the Chapters

This book is divided into eight chapters. Following the introduction, Chap. 2 analyses theoretical contributions to the debate on the causes of women's unequal pay. The chapter begins with a consideration of different forms of power before discussing Lukes' (2005) three dimensions of power, with a particular focus on the issues of non-decision-making and covert forms of power. Bradley's model of gendered power resources (Bradley 1999) is fully explained and then extended to incorporate legal power, something women have fought for using their collective power. The model then builds to discuss the division into primary and secondary power resources and how they can reinforce each other. Women's late entry into factory work is seen as decisive in weakening their access to crucial power resources.

The rest of the chapter divides between economic explanations for unequal pay, gender-based theories and the workplace effects of gendered processes. Neo-classical theory is quickly disposed of before considering the effects of segmented labour markets and their ability to foster division that works to employers' advantage (Edwards et al. 1975). While segmentation is easily exposed, it is not clear that employers are always its sole creators, union sectionalism, especially that of craft unions, may also be implicated. However, there is a weakness in its application to discrimination in that it relies too heavily on pre-market disadvantage

(Walby 1988). Theorists of patriarchy (Millet 1971; Firestone 1970) were also unconvinced, believing men were the problem, rather than capitalism. Socialist feminists could not accept such an approach and, instead, sought answers in dual systems theory (Hartmann 1981). The articulation of these processes was difficult to pin down, and it retained the problems of patriarchy's lack of explanatory power and its attendant problems of essentialism and ahistoricism (Rowbotham 1982).

Subsequently, the adoption of intersectionality, a looser conception, allows for complexity, especially in recognising that all women are not the same. It was black women that created this awareness (Crenshaw 1991). However, the tendency to a post-structural or post-modernist approach that some intersectional analysis incurs is not suitable for an industrial relations approach. Instead, this analysis adopts what may be termed a mid-level intersectional investigation (McCall 2005) and seeks to do justice to both the class and gender elements.

Chapter 3 is the first of two chapters looking at the historical processes relevant to understanding gendered workplace practices. This chapter primarily focuses on men's work organisation and pay processes and how they established their dominant position in both—relative to women. It covers the period from the introduction of printing into Britain in the late fifteenth century up to the present day. It examines the formation of the printing craft guild, its effect on the organisation of jobs in the industry and the exclusion of women. Importantly, this gender group formation established men as the dominant force and embedded it to the extent that it became invisible; class relations effectively subsumed into male class relations. The chapter continues with an analysis of the effects of the collapse of this system and how its practices were, nevertheless, carried over into capitalist work organisation. It highlights how the unions addressed divisions in the labour process that created new male segments (contrasted with female segments in Chap. 4). In particular, it exposes the heavily masculine culture of the print chapel.

The chapter also discusses the development of trade unionism, the war of attrition with employers and the development of employers' associations. The advent of national bargaining is covered and the appearance of a plethora of extra payments that had such a marked effect on the gender pay gap (GPG) subsequently. Post Second World War, employers began to gain greater traction in delivering efficiency and productivity, and this was greatly assisted by technological developments that put a strain on union relations. The resulting mergers are explained, especially the failure to breach the craft/non-craft divide until 1991. The formation of the GPMU is shown to have been long overdue and a response to industrial and political developments that was, arguably, 'too little, too late'. The chapter ends with a brief discussion of the merger that ended independent print trade unionism in Britain.

Chapter 4 explains women's entry into the industry and its contested nature. Women are shown to have acted in solidarity with men on many occasions, but where men supported women, it appears it was mostly as a bulwark, protecting men's pay and conditions (Musson 1954; Bundock 1958). Women were excluded where possible and segregated where not. Even when women were first unionised, it was into their own unions, which were generally too weak to prevent their use as cheap labour. Eventually the men's unions were forced to bring them into membership. However, they had limited direct representation, even the forerunner of the London Women's Branch was initially reliant on a male official. It is also apparent that men were able to mobilise state power (Walby 1986) to ensure that women's encroachment into men's jobs was removed after each of the world wars. Gradually, from the 1960s on, women began to gain traction within unions, if not in the workplace. Employers utilised labour market segmentation to their advantage, while male trade unionists protected their status and pay, supporting occupational segregation of women. Even the Equal Pay Act is shown to have had minimal effect. However, from the 1980s structural change within unions increased women's influence, culminating in guaranteed proportional representation (GPR) in the GPMU. But the progressive measures appearing, as they did, so late were insufficient to improve women's access to jobs and pay before industry restructuring removed large numbers from the industry. Women had started out literally invisible and were moving back into that situation in the 2000s.

Chapter 5 moves on to examine the twenty years prior to the research: the period of greatest technical change. The labour process is examined, especially as it relates to the gender divide. Job content and barriers to women's progression are revealed. Traditional factors played a part, such as men's claim to be the sole workers responsible for heavy lifting, despite evidence to the contrary. The effects of increasing flexibility are also examined. The latter was seen as implicated in the removal of women from the industry, as a response to intensified competition. Skill continued to be a key factor and the monopolisation of skill definitions in favour of men. The outcome was that women remained trapped in jobs defined as unskilled, and the link between the national agreement and job allocation (Craig et al. 1984) was shown to be decisive. Men also became subject to segmentation processes, reducing the numbers able to claim the highest pay.

Chapter 6 examines employer and trade union approaches to tackling discrimination. The existence of a gender inequality regime (Acker 2006) in relation to recruitment, promotion and training becomes apparent. Working patterns, favourable to men but problematic for women with domestic responsibilities, were seen to drive self-selection for the industry, women leaving the industry when children were young and returning at a later time. This is a clear example of an industry resistant to taking responsibility for domestic commitments (Acker 1998). The maintenance of inequality regimes is demonstrated to occur through the operation of asymmetrical power resources that 'proceduralise' responses to discrimination that, ultimately, support the status quo. The chapter moves on to consider union structural change. Despite evidence that the union was taking discrimination more seriously, women were still marginalised and so were their bargaining issues, especially pay. Structural change was insufficient, requiring women to make huge sacrifices in order to influence decision-making, and, as McBride (2001) has also found, being blamed for lack of progress. Their relative invisibility is, thus, maintained, and, again, 'proceduralisation' is implicated in weakening challenges to the gender regime.

Chapter 7 covers the issues of pay structure, determination and outcomes that are closely linked to work organisation processes. The national agreement played a key part in this analysis, supplemented by a discussion of local bargaining. Member interviews in particular provided an important element in this discussion as they revealed how members respond to processes in which they are largely passive recipients. Class relations demonstrate how the balance of power shifted considerably in favour of employers (Dickens 1997), adversely affecting women, whose concerns were only beginning to be addressed at this time of union weakness. Economic, social and political factors are acknowledged to be responsible for this development. 'Proceduralisation' is most apparent here, given the effects of pay determination processes. Walton and McKersie's seminal work (Walton and McKersie 1965) is given a gendered reading to make visible the importance of symbolic power in delegitimising both issues of particular importance to women and women as negotiators. As a result, women's issues became visible, but little substantive equality was realised.

Finally, Chap. 8 proposes an explanation for these empirical observations. Utilising the model of gendered power resources, it is argued that women came late to production work. Employers had established the key resource of economic power, while men had acquired technical, collective, sexual and physical power resources. These were sufficient to ensure gendered work organisation and, therefore, pay structures from the outset. Women became a source of cheap labour and men could secure their status against this. While examples of collusion between male workers and employers can be found, it is argued that this is not necessary given the coincidence of interest they shared (Lown 1983). The stability of technology made it difficult to shake gendered work organisation, even when technology changed. Only in the late twentieth century did women begin to marshall sufficient power to challenge their position in the industry. However, technical change, when married to political change, weakened unions at the point where they were beginning to provide greater support to women. In the event, 'proceduralisation' became a mechanism that employers and male workers, sometimes wittingly, sometimes not, utilised to maintain the status quo and retain a focus on class relations that are predominantly identified with white, male trade unionism. Asymmetrical access to power resources made this possible. Ultimately, large-scale contraction of the printing industry removed women's jobs to an even greater extent than men's, rendering them once more largely invisible in work organisation and pay processes. This book begins to explain these processes by outlining existing theories of discrimination and how an analysis of power can strengthen such explanations.

Note

1. In this book 'communications' is a loose term denoting recognisable groups of workers such as those in publishing or information technology. However, this work is also being undertaken by customers through desktop publishing, as is made clear in Chap. 5.

References

Acker, J. (1990). Hierarchies, Jobs, Bodies: A Theory of Gendered Organizations. *Gender and Society, 4*(2), 139–158.

Acker, J. (1998). The Future of 'Gender and Organizations': Connections and Boundaries. *Gender, Work and Organization, 5*(4), 195–206.

Acker, J. (2006). Inequality Regimes: Gender, Class, and Race in Organizations. *Gender and Society, 20*(4), 441–464.

Bradley, H. (1996). *Fractured Identities: Changing Patterns of Inequality*. Cambridge: Polity Press.

Bradley, H. (1999). *Gender and Power in the Workplace*. Basingstoke: Macmillan Press.

Bradley, H. (2007). *Gender*. Cambridge: Polity Press.

Bryman, A., & Bell, E. (2011). *Business Research Methods*. Oxford: Oxford University Press.

Bundock, C. J. (1958). *The National Union of Printing, Bookbinding and Paper Workers*. London: Oxford University Press.

Cockburn, C. (1983). *Brothers: Male Dominance and Technological Change*. London: Pluto Press.

Colgan, F., & Ledwith, S. (1996). Sisters Organising – Women and Their Trade Unions. In S. Ledwith & F. Colgan (Eds.), *Women in Organisations*. Basingstoke: Macmillan.

Collins, P. H. (1990). *Black Feminist Thought, Knowledge, Consciousness, and the Politics of Empowerment*. London: Routledge.

Collinson, D., Knights, D., & Collinson, M. (1990). *Managing to Discriminate*. London: Routledge.

Connell, R. W. (1987). *Gender and Power*. Stanford, CA: Stanford University Press.

Craig, C., Garnsey, E., & Rubery, J. (1984). *Payment Structures and Smaller Firms: Women's Employment in Segmented Labour Markets*. Research Paper No. 48. London: Department of Employment.

Crenshaw, K. (1991). Mapping the Margins: Intersectionality, Identity Politics and Violence Against Women of Color. *Stanford Law Review, 43*, 12–41.

Dickens, L. (1997). Gender, Race and Employment Equality in Britain: Inadequate Strategies and the Role of Industrial Relations Actors. *Industrial Relations Journal, 28*(4), 282–291.

Edwards, R., Reich, M., & Gordon, D. (1975). *Labor Market Segmentation*. Lexington, MA: D.C. Heath.

Firestone, S. (1970). *The Dialectic of Sex*. London: The Women's Press.

Foster, D., & Williams, L. (2010). The Past, Present and Future of Workplace Equality Agendas: Problems of Intersectionality in Theory and Practice. In P. Blyton, E. Heery, & P. Turnbull (Eds.), *Reassessing the Employment Relationship*. Basingstoke: Palgrave Macmillan.

Fraser, N. (2008). From Redistribution to Recognition: Dilemmas of Justice in a "Postsocialist" Age. In K. Olson (Ed.), *Adding Insult to Injury: Nancy Fraser Debates Her Critics*. London: Verso.

Gennard, J., & Bain, P. (1995). *SOGAT: A History of the Society of Graphical and Allied Trades*. London: Routledge.

Gottfried, H. (1993). Notes Towards the Development of a Liberatory Research Project. In K. Forrester & C. Thorne (Eds.), *Trade Unions and Social Research*. Aldershot: Avebury.

Harding, S. G. (1987). *Feminism and Methodology*. Indiana: Indiana University Press.

Hartmann, H. (1981). The Unhappy Marriage of Marxism and Feminism: Towards a More Progressive Union. In L. Sargent (Ed.), *Women and Revolution: A Discussion of the Unhappy Marriage of Marxism and Feminism*. Boston: South End Press.

Healy, G. (1997). The Industrial Relations of Appraisal: The Case of Teachers. *Industrial Relations Journal, 28*(3), 206–219.

Hesse-Biber, S., & Yaiser, M. (2004). *Feminist Perspectives on Social Research*. Oxford: Oxford University Press.

Holgate, J., Hebson, G., & McBride, A. (2006). Why Gender and 'Difference' Matters: A Critical Appraisal of Industrial Relations Research. *Industrial Relations Journal, 37*(4), 310–328.

Ledwith, S. (1991). *Sweet Talk, Soft Deals? – Gender and Workplace Industrial Relations*. Working Papers in Economics, Business and Management No. 35. London: Polytechnic of North London.

Ledwith, S., & Colgan, F. (2002). Tackling Gender, Diversity and Trade Union Democracy. In F. Colgan & S. Ledwith (Eds.), *Gender, Diversity and Trade Unions*. London: Routledge.

Ledwith, S., & Hansen, L. L. (2013). *Gendering and Diversifying Trade Union Leadership*. London: Routledge.

Ledwith, S., Hayes, M., Joyce, P., & Gulati, A. (1985). *Women in SOGAT '82: Report of a Research Project into the Role of Women in the Union Carried Out by an Independent Research Team from the Polytechnic of North London*. Essex: SOGAT.

Letherby, G. (2003). *Feminist Theory in Research and Practice*. Milton Keynes: Open University Press.

Lown, J. (1983). Not So Much a Factory, More a Form of Patriarchy: Gender and Class During Industrialisation. In E. Gamarnikow et al. (Eds.), *Gender, Class and Work*. London: Heinemann Educational Publishers.

Lukes, S. (2005). *Power: A Radical View*. London: Palgrave Macmillan.

McBride, A. (2001). *Gender Democracy in Trade Unions*. Aldershot: Ashgate.

McCall, L. (2005). The Complexity of Intersectionality. *Signs, 30*(3), 1771–1800.

Millet, K. (1971). *Sexual Politics*. London: Virago.

Musson, A. E. (1954). *The Typographical Association*. London: Oxford University Press.

Roe, A. (1996). John Gennard and Pater Bain – A History of the Society of Graphical and Allied Trades. *Historical Studies in Industrial Relations, 2*, 184–188.

Roe, A. (2003). *The National Agreement in the Printing Industry – The Exception to Test the Rule*. PhD Thesis Unpublished, Keele University, Staffs.

Rowbotham, S. (1982). The Trouble with 'Patriarchy'. In M. Evanston (Ed.), *The Woman Question*. London: Fontana.

Wajcman, J. (2000). Feminism Facing Industrial Relations in Britain. *British Journal of Industrial Relations, 38*(2), 183–201.

Walby, S. (1986). *Patriarchy at Work*. London: Polity.

Walby, S. (1988). *Gender Segregation at Work*. Milton Keynes: Open University Press.

Walton, R., & McKersie, R. (1965). *A Behavioral Theory of Labor Negotiations*. New York: McGraw Hill.

Yancy Martin, P. (2003). "Said and Done" Versus "Saying and Doing": Gendering Practices and Practicing Gender at Work. *Gender and Society, 17*(3), 342–366.

2

Theories of Discrimination

Discrimination can be taken as the application of some 'superficial personal characteristic … in an attempt to restrict an individual's opportunity for economic or social advancement' (Adnett 1996: 170). Gender discrimination, with which this book is concerned, has been a visible phenomenon in both capitalist and non-capitalist countries (Crompton 1997). Many theories have been developed to try and account for the persistence and pervasiveness of gender discrimination, generally divided between the economic and social (Becker 1971; Firestone 1970; Hartmann 1981; Crenshaw 1991). This book aims to synthesise arguments that cross this divide to establish how power relations within the printing industry affected women's ability to achieve equal pay. The intention is to demonstrate how power resources are gendered and permeate relationships between employers and workers but also between co-workers, especially between women and men.

Starting from the premise that the 'sex dichotomy cuts across all classes and strata' (Acker 1973: 178) and utilising the mechanism of gender inequality regimes (Acker 2006), the obvious, the less visible and the invisible elements of process and practice will be uncovered within work

T. Dawson, *Gender, Class and Power*, https://doi.org/10.1057/978-1-137-58594-3_2

organisation and pay determination to explain women's continuing economic disadvantage, even in an apparently changing labour process. The argument presented below suggests that asymmetrical access to power resources, presaged on the timing of entry to an occupation, embeds existing work organisation and pay structures through practices that create a mask of relative invisibility that renders women seemingly unimportant to production and bargaining processes.

The analysis in this book uses an intersectional, industrial relations framework. Understanding the employment relationship is at the heart of industrial relations, but this field of study, traditionally, has not reckoned with gender discrimination, in which improvements can seem glacially slow. Generally, this is motivated by a concern that such research will be divisive and detract from class analysis (Foster and Williams 2010). The aim of the discussion presented here is to demonstrate that this is an unnecessary fear. As Anna Pollert argued there are 'no ungendered class relations, and conversely, there are no gender relations without a class dimension at some level' (Pollert 1996: 646). Instead the focus of this research is on 'lived experience, which shows gender and class to be processual and dynamic … [considering] what can appear to be contradictory processes … breaking down traditional binarisms … [but without slipping into] a post-modernist "anything goes" standpoint' (Holgate et al. 2006: 325). This falls broadly within the ambit of intersectionality (Crenshaw 1991), whereby class and gender are seen to interact to create relative disadvantage for women. Class analysis provides an explanation of workers' disadvantage, but relative to this, women's hardship tends to be greater, meaning there is a need to uncover the class and gender processes responsible. McCall (2005) identified a mid-level of intersectional analysis that allows this study to avoid the dangers of post-structural analysis and retain key features of a structural, industrial relations approach.

The chapter begins with a short discussion of analyses of power before examining Lukes (2005) and Bradley (1999) in more detail. This leads to a model of power resources that will be applied in the rest of the study to issues of work organisation, employer and union approaches to equalities and pay determination. Subsequently, existing economic and gender-based explanations of discrimination are discussed: neo-classical, labour market segmentation, patriarchal, dual systems and intersectionality, extending to consider gendered workplace practices. The chapter concludes that a gendered model of power resources, coupled with segmentation analysis, enables the complex processes of gender relations that permeate workplace relationships to become visible and, thus, more open to challenge and change.

Power, Gender and Pay

Power has, in the main, not been discussed in great detail within industrial relations (Kelly 1998), whereas it is a concept that has been central to political science and sociology. Despite this lacuna, industrial relations is a field of study that draws on a number of disciplines to develop explanations of how the employment relationship works. In this study, the explanation for the persistence of pay inequality will draw on economics, sociology, history and law. All of these will be examined in the light of the power resources available to the parties to the employment relationship.

Power

Even defining power is a difficult task. Haugaard and Clegg (2009) argue that power 'is a cluster of concepts … central to the constitution of social order' (pp. 3–4) and it has a long history in the social sciences. The discussion normally commences with Hobbes *Leviathan* with its focus on coercive power. While this is probably the most common view of what power is, it is too limited to capture the nature of social relations. This form of power is often termed *power over*; a more subtle variant is *power to* or the capacity to gain advantage or realise objectives; power theory ultimately broadened to an understanding of power as diffuse (Foucault 1994). *Power over* was a preoccupation in the 1960s and 1970s with key contributions from Dahl (1957) and Bachrach and Baratz (1963). The latter were an improvement on the former because they began to extend the understanding of power to the less visible forms it might take. However, this does not negate the relevance of coercive power because its exercise 'narrows … [the] field of action' (Göhler 2009: 28) of those subjected to it. Bachrach and Baratz (1963) introduced the notion of *non-decision-making* whereby not doing something is also an exercise of power. Both approaches were subjected to criticism, not least from Lukes (2005) who saw the reliance on observable behaviour as a constraint on understanding power. He also criticised the failure of these conceptions to account for their tendency to individualise power, ignoring the 'function of collective forces and social arrangements' (p. 26).

Power to is defined as a capacity that may be exercised or latent (Göhler 2009), and if the latter, it may operate by threat alone. It can also be seen as necessary to the exercise of *power over*. This results in *power to* often being understood as presaged on the access to resources needed to exercise *power over*. It may also be understood as empowerment (Arendt 1970 cited in Göhler 2009), a problematic concept within industrial relations (Edwards 2003). However, Göhler (2009) captures a useful connotation of this form of power in his transitive/intransitive model, whereby he explains the notion of *empowerment with* as collective action:

> It is the trademark of common action or a common basis for individual action that the options to act are not arbitrary but oriented toward a community or at least framed by common values. (p. 37)

Amy Allen's concept of *power with* (Allen 2009) is also an acknowledgement of the need to account for resistance and ensure agency is not ignored. In this respect, Göhler (2009) highlights the work of Gidden's on structuration: the theory that 'society is the product of reciprocal process' (p. 30). This is a continuing process in which '[h]uman agency produces structures which simultaneously serve as the conditions for reproduction of human agency' (Clegg 1989: 139). Certainly, much gender-based theory takes this view, as does a key text for this analysis (Bradley 1999). The complementarity of structuration allows for the direct suppression of women but also for the overwhelming of their resources because of the asymmetrical structures in which power is implicated.

Lukes is also a key source for this discussion. His revised consideration of the radical view of power in 2005 provides a useful way of identifying the covert operation of power. He defines power relations as 'agents' abilities to bring about significant effects, specifically by furthering their own interests and/or affecting the interests of others, whether positively or negatively' (ibid.: 65). However, Lukes accepts that power can also operate unintentionally as a result of the inaction of the powerful to remedy any resulting injustice (ibid.: 68). He takes coercive power and non-decision-making as his first and second dimensions of power, while his third dimension is far subtler than the first two and concerns the effects of inaction, of averting opposition or even preventing grievances arising. This is often accomplished by making it appear that there is no alternative to the status quo, naturalising existing power relations (ibid.: 28). In some instances, assumptions about power holders' likely opposition to particular demands is enough, in itself, to prevent them arising (ibid.: 42–43). Neither is it absolutely necessary for there to be intentionality if it can be inferred that power holders could or should have had such awareness (ibid.: 76).

Following the advent of post-structuralism, it became fashionable to decry such analyses because they relied on 'grand narratives' such as Marxism for their salience. Lukes' model, in particular, was criticised for requiring an omnipotent view of what people's 'real' interests were if it was to be argued that they could be gulled into accepting alternatives to those interests (Clegg 1989; Haugaard 2009). However, as will be seen in what follows, in practice it is quite clear that people can be deterred or misdirected in terms of their interests. Within the employment relationship, as studied in industrial relations, it is not seriously contended that increased income or better working conditions would not be in the interests of workers. So for this analysis it is hard to see a problem with relying on Lukes' ideas. Haugaard (2009) also argues that Lukes' conception of the *power to mislead* (sic) suggests conspiracy theory. While not denying, in the analysis contained in this book, the potential for conspiracy or collusion, he ignores the possibility of coincidence of interests (Lown 1983), whereby the same outcomes serve the purposes of different, ostensibly opposed groups. At this point the concept of hegemony is relevant. Again, Lukes is found wanting because his third dimension of power is seen as an example that 'truth is somehow exempt from power' (ibid.: 248). It is a critique of false consciousness. Although it may be a fault in Lukes that he does not deal sufficiently with how the meaning of interests is derived, the Gramscian idea that 'the strength of the bourgeoisie is manifest in their lack of need to coerce' (ibid.: 249) is, perhaps, a subtler way to conceive of the power to mislead as a rationalisation of existing practices and processes. However, it would also be misleading not to recognise the continued existence of coercion in industrial relations. Since 1979 the environment for trade unions has been increasingly regulated, while all else is deregulated (Fosh et al. 1983). The latest attempt to constrain union action is particularly punitive as regards strike action (Trade Union Act 2016), while the policing of disputes was particularly coercive in the 1980s, for instance, in the miners' strike of 1984 (Phillips 2014). These coercive powers should also be seen as creating normative views of trade union legitimacy.

Fox (1974) also adopts something similar to Lukes' third dimension, resting his understanding on the concept of discretion. Workers all have some degree of discretion even those in unskilled work. For this reason the employer cannot simply coerce them. Instead employers rely on compliance of either the pragmatic or normative kind (p. 91). Many agencies may have socialised workers into acceptance of bureaucracy and into ceding control of large areas of their work to management. To the extent that management prerogative is legitimised, workers have imbibed the tenets of normative ideology. If legitimation fails, coercion may take its place. Where workers calculate that they do not have the power to resist, they adopt pragmatic compliance. Martin (1992) demonstrates, in this regard, how power operates in the routine application of unchallenged assumptions, particularly in the 'naturalising' of what constitutes interests and what is therefore appropriate for bargaining.

While the contribution of post-structuralism/modernism is important, it tends to see power as everything and everything as power and thus 'risks defining it as nothing, and nowhere' (Jenkins 2009: 147). For feminist research it is also a problem because the focus on ever-greater differences plays into the hands of current neoliberal thinking on the primacy of the individual and 'must lead away from the radical perspectives of feminism and Marxism towards, at the best, a liberal pluralism, at the worst a nihilistic individualism' (Bradley 2007: 73) and has been decried as a form of identity politics (Fraser 2008). This is not to deny the value of widening our understanding of difference but to point out the danger of patterns of discrimination and oppression disappearing if this is taken to extremes, a potential issue with intersectionality.

One area with particular utility for this analysis is bargaining power. In general, there is little theorising of negotiation practice. Walton and McKersie's seminal work (Walton and McKersie 1965) remains as valuable as ever. There have been other attempts to consider bargaining power (Fox 1974; Martin 1992), but they tend to focus on bargaining tactics as a source of power or subsume it under the general proposition of asymmetrical power under capitalist relations (Hyman 1975). There is certainly little in terms of gendered bargaining activity (for exceptions see Colling and Dickens 1989; Heery 2006). This has been partially remedied by a discussion of the role of symbolic power in preventing women achieving equal pay (Dawson 2014). It is argued that control of agendas, prioritisation of bargaining issues and the actual practice of negotiation allow men and employers to frame gender issues as 'social' rather than 'industrial'. This has the effect of delegitimising issues of particular concern to women and encouraging their early dismissal from the negotiating process with, at best, limited success in achieving women's goals for greater equality. In general the results are little more than a restatement of the law or a statement of intent.

There will be no attempt here to identify the essence of power. The issue at hand is how access to power resources provides the means to gain advantage in a never-ending contest between the parties to the employment relationship. Power and resistance and what resources, circumstances and objectives permit or drive their engagement are the elements of analysis. With this the discussion turns to power resources.

Power Resources

Power resources are the vehicle for the exercise of power (Lukes 2005) or *power to*. Jenkins (2009) considers power resources in terms of their efficacy in achieving objectives. He explains that 'stuff becomes a resource only in the context of what is done with it' (p. 145). He makes two important points about resources. First, he identifies the importance of networks, especially as a source of information. Second, he emphasises that people use power resources to solve problems, often improvising with 'whatever resources are to hand' (p. 146). This does not ignore the silences and absences at the heart of Lukes' (2005) third dimension; it suggests that these are also a form of action that must be recognised. Such analyses require careful consideration of human agency, in order to avoid imposing our own perspectives rather than those supported by the data.

The identification of power resources raises the problem of which resources should be analysed and which ignored. One of the commonly taught models of power resources is that of French and Raven (French Jr. and Raven 1960). The resources identified were coercive, reward, legitimate, expert and referent. This has clear limitations for an industrial relations analysis because it seems to focus almost entirely on the kind of resources managers are likely to possess. Only expert power and, possibly, referent power might relate to workers. Clearly it is possible to choose a large number of potential bases from which power could flow. However, this is not such a major problem if it is entirely clear on what basis these resources are chosen, how they relate to the field of study or discipline and how well they account for relationships within that field. With this in mind it is time to consider Bradley's model of gendered power resources.

Bradley's Model of Gendered Power

Bradley identified nine power resources that mainly favour men and capitalists. Each is discussed in turn with the addition of legal power resources. Bradley situates her model in a broad definition of power as 'the capacity to control patterns of social interaction' (Bradley 1999: 33). This book, however, prefers to take the analysis further by adopting the definition utilised by Lukes, above, to allow for a specific account of the insidious and hidden nature of power. These power resources are broadly explained in Bradley (1999: 34–37) but are expanded, below, to encompass more recent insights.

Economic power clearly favours capital accumulation and rewards of all kinds such as income, property, earnings and other work-related rewards. While Bradley notes that men tend to control capital, she focuses on men's greater earnings power as a key gender power resource in producing the concept of the male breadwinner and therefore power in the domestic sphere (ibid.: 34). The success in recent decades in integrating women into paid work diminished the existence of the male breadwinner model. However, the recognition that stagnating wages for many lower- and middle-income families have simultaneously given rise to the dual-earner family, especially in countries such as the UK (Karamessini 2014: 7), does not detract from the different relative wages that still tend to favour men. The continuing decline in men's wages since the global financial crisis denotes an equalising down of pay (Rubery and Rafferty 2013; ONS 2013, 2014) and potential to break down some of the gendered protectionism around pay hierarchies.

Positional power identifies the holders of relatively powerful positions, for instance, managers, supervisors, heads of households and trade union leaders. In essence this is about the possession of authority (Bradley 1999: 34) which is both a class and gender-related concept.

Technical power is crucial for the account that follows. It refers to the possession of knowledge, expertise and competence, especially in relation to machinery and latterly computer technology. Bradley notes that Cockburn (1983) was the key exponent of the theory that the ability to assign gendered identities to new technology was a source of power (ibid.: 34).

Fox (1974) argues that discretion is a key source of power for both craftworkers and professionals. Workers will seek to enhance their discretion in order to get greater control of the job and may even try to constrain management through gaining agreement to rules that limit management's discretion. In this thesis, technical power is related to the key issue of skill definitions, how and by whom they are controlled and the discretion available to workers. Changes to the labour process suggest scope for women to enter these, previously closed, occupations. It is the failure of this expectation that is the key question for this analysis.

Physical power can be taken on one level to relate to male violence. But Bradley also explains that physical power resources enabled men to 'legitimate their dominance at work' (ibid.: 34), although this is now under threat from new technology. For the purposes of this book, physical power will be used to explain how men have used physical strength to control access to the more lucrative work, becoming almost a proxy for skill (Cockburn 1983).

Symbolic power is a crucial element in this model as it relates to control of meanings, values, definitions and rules. Definitions of skill and negotiating practices are important in this regard. Symbolic power also pervades trade union democracy through the control of agendas and through 'conversational devices to marginalise women's contributions' (Bradley 1999: 35).

Collective power involves any mobilisation of groups to pursue common aims. Political mobilisation has benefited both men and women at different times, for example, through trade unions and pressure groups. Conversely, informal workplace networks may operate to marginalise women. The other aspect of this power resource is through the collective power of trade unions. While historically this has tended to benefit men, more recently women's self-organisation has resulted in at least some improvement in their representation within trade union democratic structures (Colling and Dickens 2001). However, collective power is also available to capital in achieving political ends that support capital accumulation.

Personal power resources capture the notion of utilising personal characteristics and know-how to improve or maintain positional power. This may include resources such as 'strength of character, knowledge, ability to get on with people, charisma, experience, articulacy' (ibid.: 35) as well as interpersonal skills. Bradley also notes that women may utilise this resource to gain power over men in the home or at work. The latter may be useful in countering male domination or may help women achieve positional power.

This resource captures the features of internal politics that are so important in negotiating a way through organisations. Men tend to have a better idea of how the system works, the 'rules of the game' (Bachrach and Baratz 1963), especially in institutions such as trade unions, and these features explain how they can continue to remain powerful and resist women's encroachment even following structural change. Personal power and positional power may also be linked in terms of male managers appearing to be all-powerful as far as women are concerned, and therefore women's anticipation of discrimination may avert challenges to inequality regimes.

Sexual power resources are linked to personal power resources but in this instance have a special resonance in terms of sexual harassment. However, Bradley notes that they may have utility for women in gaining power over men and even to gain economic advantages (Bradley 1999: 35).

Sexual harassment has an important role in resisting women's encroachment into traditionally male-dominated areas (Collinson and Collinson 1996). In addition, it may provide a way for women to make their lives easier by conforming to assumptions about women's behaviour. For instance, women in manufacturing industry are seen as trespassing on men's turf (Pollert 1983), but by playing up their feminine side when at work, they may lessen the hostility that can come their way. The combination of collective and sexual power resources may also provide scope for women to challenge the structural constraints within their trade unions, while for men it has been decisive in capturing positions of power in the union and relative power at work.

Domestic power resources relate to 'control of household goods and materials'. The idea is that this provides a source of power for women in controlling subsistence and comfort. There is no evidence presented in this research of this nature because the research design was unable to capture such a notion. Domestic power resources indirectly benefit employers in justifying their lack of investment in women's training and in women's low pay. They are also crucial in the expansion of the model, below, because they indicate that women's domestic role caused them to enter the workplace in significant numbers later than men, thereby ceding ground to men in the acquisition of key technical and collective power resources.

Extending the Model

Although Bradley's model is comprehensive, it has missed an important resource that has had growing significance in the last 40 years. Discrimination law has grown substantially. This was particularly so following the election of Tony Blair's New Labour government in 1997. Albeit the legislation they enacted was usually enacted in a minimalist way as 'light touch' regulation, it has still provided greater scope for women to challenge unequal treatment. It was also New Labour that finally consolidated the many statutes and regulations into the Equality Act 2010 with potential, as yet unrealised, to deliver greater equality for all disadvantaged groups. Therefore, the model as used in this analysis incorporates:

Legal power resources as a means to theorise women's workplace resistance in greater detail. These resources would not exist without women's political mobilisation to campaign for change and recognition of women's particular interests. Neither would public policy more generally have been altered without this resource. So, collective power resources were needed in order to provide legal power resources. The limitations of legal power resources are acknowledged, given employers' ability to resist them. However, they do provide some scope for women to resist subordination, and, where there are large numbers of women who can also act collectively, results can be better (Thornley 2006).

Having established which resources are to be utilised, the analysis revealed a system of reinforcement operating between the power resources. Therefore, the model has been extended to explain how the power resources interact (Fig. 2.1).

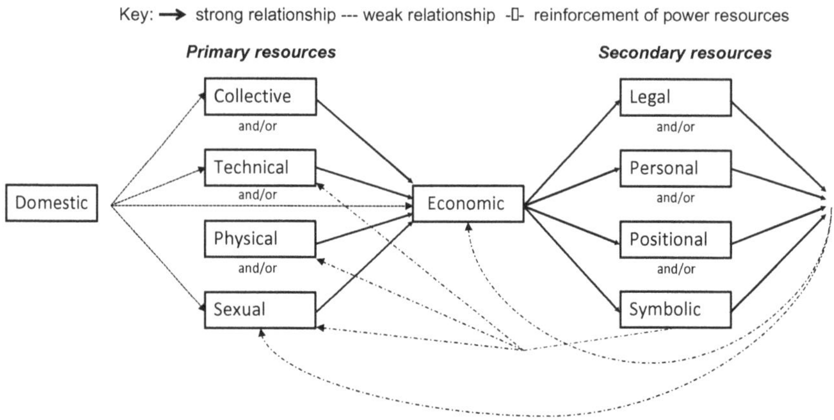

Fig. 2.1 Model of gendered power resources

Within the employment relationship economic power is seen as the main goal because it results in greater wealth, status and higher living standards. Also economic power provides a key resource within capitalism, where capital accumulation is a key factor. This is placed in the centre to indicate its status. The other power resources can then be seen as either primary (left-hand side) or secondary (right-hand side). Primary power resources are those that deliver people into particular positions when they enter the industry. Collective, technical, physical and sexual power resources provide, for example, higher positions based on skill and ability to do heavy lifting, but they also provide the scope for gendered collective organisation. The heavy black arrow indicates a relatively strong relationship to economic power. Domestic power resources are placed further to the left than the other primary resources and link to the other primary resources, as well as economic power with a dotted line, indicating that those with such power are likely to enter the workplace later and with less access to primary resources because these positions have already been colonised by those arriving earlier.

Having acquired primary power resources, secondary power resources also become available: legal, personal, positional and symbolic power. Again, those entering late may acquire some measure of these resources, but they will be weakened by the later acquisition of primary resources. Finally, secondary resources act to further strengthen primary resources. In this way, those who arrive late to the party are likely to always find themselves at a disadvantage. Even though they have power, and the agency to use it, in relative terms it is weaker than those who arrived earlier and attempts to utilise it will tend to be overwhelmed until a critical mass, in terms of numbers, pushes the situation towards a tipping point where existing relationships will come into question and may need to be re-negotiated. Shocks to the labour process could potentially provoke such change.

Women are those who tended to arrive late and so have been at a consistent disadvantage ever since, although there has been some weakening of the hold on power of men and employers. Men already dominated positions within collective organisation when women arrived. Employers and their agents, managers, holding the economic and positional powers of hire and fire, were always in the ascendant in relation to women (and all) workers. Women's sexual, collective and legal powers were the key resources at their disposal, and despite some success, their ability to transform their positions was severely limited. Exactly how this happened is the subject of the rest of this book.

Employers' trajectory through the model is slightly different. They would tend to dominate the centre through economic power. The left-hand column may not be as relevant if they acquired this position through inheritance. However, technical power is important in relation to the claims management make to their 'right to manage' (Storey 1983)—managerial expertise and experience. These would also give rise to positional and personal power resources, in particular. Historically, apprenticeship routes into becoming a master could provide greater access to power resources (see Chap. 3).

The effects of differential access to power resources will be linked to the following theories of discrimination to demonstrate how they might operate to increase women's disadvantage at work. They are divided into economic and gender-based theories.

Economic Theories of Discrimination

Neo-classical Explanations

The starting point for economic theories are the neo-classical explanations of discrimination. The initial response of neo-classical economists was to claim that sex and race discrimination did not exist in the labour market. As adherents to the logic of the market, they could not see how supply and demand could be affected by ascriptive characteristics, unless they were external to the market. According to this view, workers' pay was commensurate with their marginal productivity, or to put it another way, workers were paid what they were worth under the conditions of a perfectly competitive market. Workers looked to maximise utility, and employers looked to maximise profit (Adnett 1996). Worth was rationally calculated based on the skills and productivity of the worker—'a reduction of all social relationships to exchange relations' (Gintis 1987: 70).

This meant women were paid less because they were worth less. But the neo-classicists had overlooked the idealistic nature of their theories: markets were not perfectly competitive and rational judgement was scarce (Wootton 1955; Dunlop 1966; Gintis 1987; Sawyer 1989; Phelps Brown 1977). Labour was treated as any other factor of production, ignoring its difference 'in being human' (Sawyer 1989: 81). Instead, critical analysis revealed tolerances in the system that tended to produce diverse outcomes and allow for artificiality to creep into wage structures, undermining the mechanisms of supply and demand (Phelps Brown 1977). Such tolerances arose because firms were constrained by social factors (Brown and Nolan 1988) such as the embeddedness of existing wage structures, work organisation and related practices, the nature of technology and, particularly, the resistance to change—from both workers and managers (Craig et al. 1984).

Further problems arose from the claim that for discrimination to exist, employers would have to have a 'taste' (preference) for discrimination, which, if exercised, ensured employers would be the losers because they would fail to employ the best workers, while their non-discriminating competitors would benefit. Because of this discrimination should be competed away (Becker 1971). The continued existence of the gender pay gap undermined this argument.

As the pay processes' argument failed, neo-classicists turned to examine work organisation, through family specialisation and human capital theories. In essence, these theories are meant to explain the clustering of women either in particular occupations (horizontal occupational segregation) or in the lowest positions in the hierarchy (vertical occupational segregation) (Beechey 1986). Segregation is ubiquitous, occurring in capitalist and non-capitalist countries and appears resistant to change (Barrett 1980; Crompton 1997; Hakim 1979). Both theories claim that domestic labour and caring responsibilities account for women's low attachment to work and/or their failure to acquire skills and knowledge. Therefore, women end up in low-paid jobs. However, even some adherents to this view accepted that anticipated discrimination may play a part in turning women away from alternative jobs (Mincer and Polachek 1980, originally 1974). Of course, these explanations are predicated on married women generally with children, ignoring those who are childless. Further criticisms include the assumption that greater human capital equates with higher marginal productivity, ignoring questions of effort expended and the use of education as a screening device (Sawyer 1989). Evidence also demonstrated that women were not penalised less for periods out of the workplace and nor was their continuity of employment linked to occupational segregation (England 1982). Of course, the most damning evidence is the huge growth of women in tertiary education and girls' superior results in school (Kirton and Greene 2016). Allied with statistical discrimination, a theory based on accessing reliable information to evaluate the expected productivity of different groups, women were seen as worth less because they were expected to perform worse than men (Adnett 1996). However, if this were to hold true, a tight labour market should lead to discrimination being competed away.

The existence of internal labour markets (ILMs) has been an invaluable concept in industrial relations research. Within an ILM decisions are mostly administrative, rather than market driven, thus enabling employers to access firm-specific skills and avoid opportunistic bargaining (Rubery 1994). However, in some circumstances, it may be that industry-specific skills affect work organisation in a similar way, albeit without necessarily avoiding opportunistic bargaining. Developing from the ILM was the simplest form of segmentation theory: dual labour market (DLM) theory (Sawyer 1989; Doeringer and Piore 1985). The basic division is between primary and secondary labour markets (PLMs and SLMs), the former very similar to an ILM. The characteristics of a DLM include divisions into higher- and lower-paid sectors, restricted mobility between PLMs and SLMs, higher-paid jobs benefiting from promotion opportunities not available to lower-paid sections and higher-paid jobs having employment continuity while lower-paid ones have high turnover. SLMs provide few opportunities for promotion, to the extent that there are often barriers preventing movement into the PLM (Barron and Norris 1991 [previously 1976]). Therefore, 'workers in internal labour markets have some rights and privileges which others do not have' (Sawyer 1989: 89). This generally leaves workers in the SLM subject to the vagaries of competition.

Women's employment was linked to the SLM through their supposed characteristics of 'dispensability, clearly visible social differences, little interest in acquiring training, low economism and lack of solidarity

(Barron and Norris 1991 [originally 1976]). These assumptions ignored women's lack of bargaining power (Purcell 1979), especially where working was not a choice. Walby (1988) observed that it also emphasised pre-entry discrimination 'unmodified by the workings of the labour market' (p. 18), which relied on belief systems to provide the explanatory base. Beechey (1986) argued that DLMs could not explain why particular jobs have been designated 'women's work', focused too much on manufacturing industry and made it difficult to account for primarily female sectors of employment, such as clerical work. Moreover, disadvantaged groups might stay loyal to a company even when their pay was low, suggesting motivation was unnecessary (Craig et al. 1984). Women were consequently caught in a 'self-reinforcing process … [whereby they were aware] of the relatively low pay in jobs accessible to them and of their relatively weak bargaining position [which] shapes their expectations' (ibid.: 89).

A useful extension of the DLM thesis was its division into subtler gradations. Loveridge and Mok (1979) proposed that there were both primary internal/external labour markets and secondary internal/external labour markets with their own characteristics. In this regard industry-specific skills, with standardised wages, could conform to the idea of a primary external labour market (PELM) and elements of job security, and relatively superior wages for secondary market workers conform to a secondary ILM (SILM).

Neo-classical theories privilege economic power, only the market and capital accumulation matters, and chime with current concerns over 'neoliberalism' (Thompson 2011). This supports the positional and personal power resources of managers serving shareholder interests. Overall the reception of these explanatory characteristics was hostile, leading to more complex non-orthodox approaches.

Alternative Economic Explanations

The limitations of neo-classical explanations gave rise to alternatives that expressly considered factors ignored in orthodox theory, including power relations, history, social and political developments as well as labour market realities and custom. These alternatives included a consideration of how the labour market becomes segmented and the effect this has on different groups of workers.

In contrast to the orthodox approach, class-based models stress the role of the market in perpetuating and exacerbating existing inequalities (Sawyer 1989). Refuting the idea that exchange relations are at the heart of wage determination, the theory proposes that the split between labour and labour power is the key element leading to the exploitation over which the struggle to allocate profit and wages takes place (Gintis 1987). Marxist theorists situate discrimination within this framework where power relations between capital and labour are unequal and employers aim to overcome threats to their control posed by class struggle. But while there is some accommodation of the idea of patriarchy in this conception, it is usually restricted to pre-market discrimination that is subsequently reinforced by the market mechanism (Sawyer 1989), rather underplaying the role of both the labour market and internal labour markets (Gordon 1972; Edwards et al. 1975). In essence, the argument was that employers utilised a 'divide and rule' strategy based on pre-existing divisions that was exacerbated by bureaucratic divisions of an increasingly homogenous workforce. Workers in the higher strata would see those beneath them as a threat and those lower down would increasingly accept their poor wages and conditions (Gordon 1972). Walby (1988) criticised both the periodisation adopted and the lack of any theorisation of the pre-existing divisions.

A further explanation of divisions of labour came from Braverman (1974) who saw them as a way of reducing skill in a job, both cheapening it and facilitating employer control of workers. This has drawn attention to developments in the labour process and issues around technology, skill, segregation, desegregation and resegregation (Cockburn 1988), as well as polarisation of skills (Gallie 1991). Central to these discussions are questions around skill definitions that designate 'women's work' as unskilled even where, objectively, this is not sustainable (Beechey 1986; Phillips and Taylor 1980). However, Braverman was criticised for limiting his explanation of gender inequality to the interaction of capital and the family (Walby 1988) and for ignoring worker resistance (Beechey and Perkins 1987). These arguments also downplayed the fact that segmentation may occur through 'strategies of closure, [which] fragments and splinters a putative working class' (Crompton and Sanderson 1990: 41). Historically, such closure occurred in the exclusion of women from the old craft guilds and was carried into industrialisation in trade union organisation, arguably because of its benefits to men as much as employers (Barrett 1980; Hartmann 1979; Munro 1999).

Meanwhile if segmentation is 'basic to the current stage of capitalist development' (Sawyer 1989: 114), it suggests there will always be some groups who lose out. However, it does not presuppose that these must be gender-based, as Thornley (1996) indicated.

Some final comments on the segmentation approach also raise concerns over how far it can provide an explanation for discrimination. For instance, segmentation is predicated on fear of wage competition, but ILMs deal with non-competing groups, and the idea of a SILM suggests that even workers in the secondary labour market might retain relative advantage and be more likely to be unionised, relative to those exposed to the external labour market. They would still remain relatively disadvantaged compared to those in a PELM, but their relative advantage over those in the external labour market may encourage quiescence. For employers, this may limit wage competition but is useful where trade union organisation is strong and where industry structure provides important but more limited benefits to the employer.

Men, as the higher-paid segment, may fear women as cheap substitutes. While some evidence for women's disposability exists (Bruegel 1979), there is, arguably, limited evidence for substitution given the existence of occupational segregation (Collinson et al. 1990). However, it has been more commonly argued that sex typing of jobs operates a normative effect 'even when supply and price conditions change' (McEwen Scott 1994: 160), favouring existing occupational segregation. For instance, even where there are skills' shortages employing women is rarely seen as a solution, possibly because 'segmentation of labour markets means that the various groups are not seen … as interchangeable' (Dickens 2005: 189). Here segmentation works in favour of men and not employers. There may be an exception to this when technology changes sufficiently to undermine men's existing skills (Hacker 1990) and women are able to act as substitutes.

These theories foreground the asymmetrical power relations inherent in the employment relationship and the logic of sexual, collective power resources. However, there is not enough recognition within them of the gendered nature of power resources. The following section corrects this.

Gender-Based Theories of Discrimination

Attempts to find gender-based explanations for discrimination have ranged from efforts to mirror Marxist theories of class exploitation, through to more subtle discussions of intersectionality. There is no overarching explanation; instead it is more 'a network of insights and arguments about connections' (Connell 1985: 261), which continue to be developed. Each strand can be seen as necessary to the developing body of work on discrimination.

The arrival of 'second wave' feminism led to attempts at a comprehensive explanation for women's continuing disadvantage. Patriarchy foregrounded power relations within the family in creating gendered hierarchies. Millet (1971) saw patriarchy as essentially about economic reward, but Firestone (1970) instead considered what is constitutive of labour. For her and other radical feminists, the original division of labour was that between production and reproduction, with women responsible for the latter, the basis of their oppression. Although important, these arguments were criticised for biological determinism, a generalised view of power relationships, failure to fully discuss the interaction of patriarchy with capitalism and ahistoricism (Beechey 1986; Hartmann 1981; Rowbotham 1982). Even Millet (1971) never claimed patriarchy was a theory. Instead theorists moved on to consider dual systems, bringing class and patriarchy together.

A key element of dual systems theory was the understanding that both managers and male workers could benefit from women's disadvantage. The sex-blind nature of class analysis was insufficient to explain 'why women are oppressed as women' (Hartmann 1981: 5). Hartmann argued that patriarchy had a material, rather than ideological, base in men's control of women's labour power and that male exclusion of women from certain occupations, especially through union organisation, played a crucial role. However, segregation was also visible in non-union environments (Cockburn 1988), although Walby (1986) took the contrary view, arguing that the direction of causality was reversed; instead women were forced back into the home, and their position in the family was adversely affected. But this really only demonstrated 'the contrast … between "market" and "non-market" workers, rather than male and female' (Crompton and Sanderson 1990). Walby also criticised dual systems for seeing patriarchy as functional for capital and underestimating the tensions between the two systems (Walby 1986: 89). Her alternative six, relatively autonomous, structures of patriarchy with domestic labour as a mode of production (Walby 1986) failed to specify 'a material base and dynamics of change for patriarchy that were clearly separable from those of the mode of production' (Acker 1989: 236) and not essentialist or transhistoric. Young (1981) was also critical that separate systems left capitalism untouched by patriarchy, but her choice of job segregation was untenable given that segregation was itself a manifestation of discrimination, rather than its cause (Walby 1988). Despite its limitations, dual systems drew attention to the creation of hierarchies. Their pervasiveness is evident in research demonstrating how workers will even create one where it does not formally exist (see, e.g., Munro 1999).

The role of domestic labour has been central to analysis of the interaction between women's paid and unpaid labour for decades (Seccombe 1974; Smith 1978; Gardiner 1975). It was argued that men and employers colluded over the use of women's labour for mutual benefit—men through freedom from domestic chores and employers through cheap labour (Cockburn 1991). Walby claimed exclusion from high-paid work was the key factor (Walby 1990), but Witz (1992) demonstrated greater complexity whereby exclusion, segregation and family employment all co-existed. While collusion was alleged, it was just as likely to be a coincidence of interests (Lown 1983). This was effected by organising work so that employers gained women's cheap labour but without upsetting the patriarchal order in the home. The resulting concepts of the family wage and the male breadwinner had considerable impact on designating appropriate jobs for each sex, although they have weakened more recently (Bruegel and Perrons 1995). In fact more women work now, including full-time, whether or not they have children. The stagnating wages witnessed in recent decades resulted in dual-earner families and a double burden for women, so that even where men do assist in housework, the bulk of the responsibility continues to lie with women (Karamessini 2014; Speakman and Marchington 1999; Windebank 2001). Concern is now focused on whether women can be forced back into the home during recessions (Rubery 2014).

Finally, there is the issue of invisibility. Throughout history women were overlooked (Rowbotham 1973). At each juncture that women challenged their subordination they were soon forgotten, whether it was after the success of the suffrage movement or within 'second wave' feminism, women's protests were ignored or marginalised (Coote and Campbell 1987) and seen as irrelevant. Even in recent times women's invisibility has been implicated in undermining their worth (Thornley 2002; Dawson 2010, 2014).

While the notion of patriarchy alone has descriptive utility, it is lacking in subtlety. Especially over time, discrimination has become more covert as women's campaigning has made it less acceptable. Power is more diffuse than is allowed here, there is insufficient recognition of women's agency and how men outflanked them in gaining access to primary power resources, allowing them to block women's access to jobs, albeit in response to the economic power of employers. Intersectionality and the concept of inequality regimes provide a more nuanced route to analysing employers' policies and trade union practice and follow in the next section.

Gendered Practices and Policies: Workplace and Union

Recruitment, selection, promotion and training work closely to create gender inequality regimes. Even where women have gained similar levels of qualifications and experience to men, occupational segregation may not decrease; employers are never sex-blind in their selection (Cockburn 1991) and fail to recruit women in non-traditional areas. Informal mechanisms and networks are influential, and employers prefer their cheapness to alternative, more transparent mechanisms (Grieco and Whipp 1986; Burchell et al. 1994; Dickens 2005). Unions may also operate as employment agencies (PPITB 1982). All these practices tend to replicate the divisions in existing workforces. Furthermore, where there is a transferable skills agreement 'to abide by a common pay system', this may restrict movement because there are few advantages for the employee in moving (Rubery 1994: 52), especially women in a secondary ILM (Burchell et al. 1994).

Even the increased presence of women managers is constrained by the masculine hegemony in organisations (Connell 1983). Such hegemony may reveal itself in recruitment and promotion decisions around suitability or acceptability, attitudes to personal attributes such as social and tacit skills and in which gender is embedded (Dickens 2005; Harris and Morris 1994; Cockburn 1991; Keep 2005). Even low-level promotions may result in men becoming supervisors over women whose work they little understand (Craig et al. 1984). Separate entry points can present an absolute bar to promotion. Thus normative values of skill predominate even where, for instance, apprenticeship-based skills are no longer relevant, reducing women's opportunities.

Acquiring skills presumes access to training, something UK employers are reluctant to provide, especially for women from whom they fear they will not see sufficient return (Dickens 2005; Keep and Rainbird 1995). Short periods of training have been linked to low pay (Dex et al. 1994), and the advent of NVQs seems not to have made much impact. Men tend to expect training and this feeds their 'sense of owning the organisation' (Cockburn 1991: 46), albeit that employers may fear bestowing bargaining power on the skilled (Rubery 1994). Instead, employers prefer to benefit from women's tacit skills and general education, rather than incur training costs (Craig et al. 1984).

Sexual harassment is also a problem for workers. It is a clear utilisation of sexual and physical power resources. It prevents women entering non-traditional jobs (Collinson and Collinson 1996), among others, by designating their sexuality as part of their labour. Women may 'quit jobs … [be] transferred or demoted [or] lose their jobs' (Stanko 1988: 92–93) because of harassment; they are intimidated into knowing their place through awareness of 'the high social cost' of transgression' (Cockburn 1988: 40). Conversely, male sexuality may flourish through the bonding exercise of 'sexual banter and exchanges …. [which may be] … at the expense of women' (Cockburn 1991: 148).

Equal opportunities policies (EOPs) routinely fail women who pursue demands for fairness, while employers are primarily interested in the 'bottom line' (Cockburn 1991), as demonstrated in the business case for equality (Dickens 2005; Noon 2007). These limitations undermine the liberal approach of meritocracy, inherent in EOPs, and fail to understand the structural sources of social inequality (Jewson and Mason 1992, originally 1986). While the weaknesses of the implementation gap are well known, the procedural effects on outcomes are less well-documented. An argument pursued in this book is taken from industrial relations history. In 1968 the Donovan Commission saw greater proceduralisation as a route to gaining control over anarchic workplace relations. Since then procedures have come to dominate workplace relations, often emanating from good practice models suggested by government bodies such as ACAS and the EHRC (acas.org.uk; ehrc.org.uk). Subsequently, the notion of procedural justice (Bingham 2016) has taken root, establishing the idea that workers will tend to accept even decisions they do not like, as long as they are seen to have gone through transparent, consistent procedures. This raises questions about how far procedures are likely to assist in naturalising the status quo, as with Lukes' (2005) more subtle third dimension of power. Such processes would clearly privilege personal and positional power resources both within workplaces and union organisation.

Moreover, the weakness of 'light touch' regulation has restricted the law's ability to discipline employers (Smith and Morton 2006). This has left equality law operating in tension with prevailing social policy, whereby '[a]nti-discrimination legislation exists as something of an island in a sea of general deregulationary measures which run counter to equality' (Colling and Dickens 2001: 285). Even this has fallen foul of 'austerity' with the Coalition government's weakening of legal rights in the Enterprise and Regulatory Reform Act 2013, making this route less likely to produce equal outcomes, although the recent successful legal challenge to the imposition of tribunal fees may partially alleviate this situation (unison.org.uk, July 2017).

Trade unions have increasingly become involved in tackling gender inequalities but may also be sites of discrimination (Boston 1987; Drake 1920). For instance, union exclusion and segregation of women were implicated in embedding occupational segregation and limiting equal pay (Hartmann 1979; Walby 1986). However, increasing regulation has severely weakened unions at a time when they began to acknowledge discrimination internally, as well as externally (Fosh et al. 1983; Ledwith and Colgan 2002).

Nevertheless, internal union restructuring has provided women with influence on decision-making bodies (Parker 2009; TUC 2016). But for women union officials, there has been the problem of fitting themselves into a masculine culture of long hours under the assumption that there are no domestic responsibilities (Wajcman 2000; Kirton and Healy 1999; Kirton 1999). The numbers of women officials remain limited, given officials tend to come from high-paid sectors, and they may face 'inequity, sexism and discrimination' (Briskin 2006: 369; Heery and Kelly 1989; Hyman 1994). The reification of masculinity may also preclude women from becoming negotiators (Willis 1977). This raises the issue of relative privilege and how it makes discriminatory outcomes invisible. However, privilege should not be seen in universal terms but as a tool to examine spheres where groups possess privilege and where they do not (Kimmel and Coston 2017). This is helpful in acknowledging the contradictions that come from gender and class intersectionality and is particularly evident in union organisation.

Unions have three methods of operation: mutual insurance, legal enactment and collective bargaining. The first two have affected women through increasing prosecution of discrimination claims (Colling and Dickens 2001) and in campaigning for equality legislation and policy. The latter manifests as equality bargaining, within which equal pay bargaining is limited (Heery 2006). The effect of bargaining processes is a crucial issue. In practice it has been argued that collective bargaining reinforces inequality by pursuing the interests of hegemonic groups (Dickens 1997). While multi-employer bargaining has been regarded as helpful in pursuing equality, it has also been recognised that it might maintain and reinforce gender differentials (Rubery and Fagan 1995). Ledwith (1991) goes so far as to claim local bargaining can lead to 'sweetheart deals' between male representatives and management to the benefit of men, in contrast to Craig et al. (1984) who see local bargaining as important for the low paid. Actual face-to-face negotiating behaviour is also an important factor. Walton and McKersie (1965) discussed the ways in which tacit understandings between the parties, the variety of tactics pursued and the need to satisfy the demands of intra-organisational constituencies can affect the success of bargaining agendas. While not devised as a method of examining equality bargaining, it reveals the effects of symbolic power on unequal outcomes (Dawson 2014).

Despite increasing equality bargaining, it has been found to be wide but shallow, mostly confined to procedural agreements (Colling and Dickens 2001; Heery 2006). It has also been less common in manufacturing industry. Furthermore, the existence of women officials appears to have only limited effect. Heery suggests that for equality bargaining to flourish, the agenda must allow groups to identify their specific interests, negotiators must be willing to pursue them and the institutional arrangements for collective bargaining and public policy must provide both opportunities for and pressure to negotiate equality (pp. 538–540). However, employers can effectively avoid equality bargaining by questioning its legitimacy, designating such issues as 'social' (Acker 1998: 200), an exercise in symbolic power.

Work organisation and skill designation are relevant to the pursuit of equality bargaining. Skill is socially constructed usually to the detriment of women (Phillips and Taylor 1980). In craft occupations skill has been used to legitimate the pay structure, although it might benefit some low-paid workers relative to others (Barrett 1980; Craig et al. 1984; Crompton and Sanderson 1990). Skill often designates who gets extra payments (Craig et al. 1984). However, informally learned skills are often unrecognised and can be invisible when identifying who should receive higher pay, while technology can undermine these processes and the tactics of negotiation related to them. For instance, technology may undermine the process of wage leadership, which focuses on the high paid (Cockburn 1983), where it polarises skill definitions (Gallie 1991) and may leave an ever-smaller elite as the only beneficiaries. This may be painful for those who rely on the existing labour process to legitimate their masculinity (Cockburn 1994), to justify authority roles and thus positional power and may explain the resegregation that often occurs after periods of change (Cockburn 1988; Rubery and Fagan 1995). The marginalisation of women's skills can thus contribute to their invisibility or may signify ingrained attitudes. Furthermore, custom, prestige and appropriate living standards are claims that have been used to preserve superior wages for men (Wootton 1955; Acker 1973; Phelps Brown 1977; Brown and Nolan 1988), and any challenges will likely be vehemently rebuffed.

Symbolic, personal and positional power resources are particularly important in defining who benefits from practice and policy and in embedding gender relations. Legal power has provided only limited redress. The evidence above leaves no doubt about the robustness of gender inequality regimes.

Conclusions

This chapter has provided a modified model of power resources that will be useful in uncovering how the mechanisms of discrimination operate. It has considered the role of economic theories and gender-based theories in explaining discrimination. While acknowledging post-structural and post-modern arguments on the nature of power, their tendency to privilege the individual limits their scope for considering group disadvantage and the collective relations that often operate within the employment relationship. Instead the model of power resources is used to consider the group nature of structure and agency within a detailed industrial case study.

Although this book does not shy away from analysing the internal workings of trade unions and the resulting discrimination, the power resource model reveals that much gendered practice is in response to the economic and positional power of employers—that is, class relations. The importance of understanding the reactionary nature of trade unions should not be underestimated, especially given the hostile climate they have inhabited for almost four decades. Unions may also be prisoners of their history, clinging to approaches that have worked in the past, especially when relatively privileged sections of membership dominate. However, part of the argument advanced in the following chapters is that this has hamstrung them in a period of rapid change, to the extent that it is no longer just women that are losing out. There is a need to make power relations visible and problematise them in order to make headway.

While neo-classical theories privilege the market and economic power, the alternative, radical theories recognise the inherent asymmetry in the power of the parties to the employment relationship—economic and positional power in particular. Their limitation is in seeing discrimination as external to the labour market. This restricts understanding that 'divide and rule' is, often inadvertently, supported by workers clinging to sectional interests, instead of building solidarity across work groups. Men do not wish to acknowledge their relative power in comparison to their women co-workers and thus their relative privilege. Rendering these processes explicit makes it harder to ignore them and may provoke debate on change.

The gender-based theories have both strengths and weaknesses and do not constitute an over-arching theory. However, they do reveal the mechanisms through which power resources operate and help to establish, in the detailed account that follows, how these resources interact. In essence, they demonstrate how skill (technical power) subject to men's organisation (sexual and collective power) provides bargaining power that gives access to relative economic power in building a pay hierarchy. Once established the secondary power resources, especially symbolic, personal and positional power, help to embed the hierarchy. While women have some access to all these power resources, their late arrival in the industry weakens their effect. However, their sexual and collective power helped them organise and campaign to achieve legal power and, subsequently, greater positional power in the union. These resources have been insufficient to effect transformation in the union or the industry. Exactly how this situation arose is the subject of the rest of this book, beginning with an examination of historical processes of power formation.

References

Acker, J. (1973). Women and Social Stratification: A Case of Intellectual Sexism. In J. Huber (Ed.), *Changing Women in a Changing Society*. Chicago and London: University of Chicago Press.

Acker, J. (1989). The Problem with Patriarchy. *Sociology, 23*(2), 235–240.

Acker, J. (1998). The Future of 'Gender and Organizations': Connections and Boundaries. *Gender, Work and Organization, 5*(4), 195–206.

Acker, J. (2006). Inequality Regimes: Gender, Class, and Race in Organizations. *Gender and Society, 20*(4), 441–464.

Adnett, N. (1996). *Labour Market Policy*. Harlow: Pearson Education.

Allen, A. (2009). Gender and Power. In S. R. Clegg & M. Haugaard (Eds.), *The Sage Handbook of Power*. London: Sage Publications Ltd.

Bachrach, P., & Baratz, M. (1963). Decisions and Nondecisions: An Analytical Framework. *American Political Science Review, 57*, 641–651.

Barrett, M. (1980). *Women's Oppression Today*. London: Verso.

Barron, R. D., & Norris, G. M. (1991 [originally 1976]). Sexual Divisions and the Dual Labour Market. In D. Leonard & S. Allen (Eds.), *Sexual Divisions Revisited*. London: Macmillan.

Becker, G. S. (1971). *The Economics of Discrimination*. Chicago and London: University of Chicago Press.

Beechey, V. (1986). Women and Employment in Contemporary Britain. In *Women in Britain Today*. Milton Keynes: Open University Press.

Beechey, V., & Perkins, T. (1987). *A Matter of Hours: Women, Part-Time Work and the Labour Market*. Cambridge: Polity Press.

Bingham, C. (2016). *Employment Relations: Fairness & Trust in the Workplace*. London: Sage Publications Ltd.

Boston, S. (1987). *Women Workers and the Trade Unions*. London: Lawrence and Wishart.

Bradley, H. (1999). *Gender and Power in the Workplace*. Basingstoke: Macmillan Press.

Bradley, H. (2007). *Gender*. Cambridge: Polity Press.

Braverman, H. (1974). *Labor and Monopoly Capital: The Degradation of Work in the Twentieth Century*. New York: Monthly Review Press.

Briskin, L. (2006). Victimisation and Agency: The Social Construction of Union Women's Leadership. *Industrial Relations Journal, 37*(4), 359–378.

Brown, W., & Nolan, P. (1988). Wages and Labour Productivity: The Contribution of Industrial Relations Research to the Understanding of Pay Determination. *British Journal of Industrial Relations, XXVI*(3), 339–361.

Bruegel, I. (1979). Women as a Reserve Army of Labour: Note on Recent British Experience. *Feminist Review, 3*, 12–23.

Bruegel, I., & Perrons, D. (1995). Where Do the Costs of Unequal Treatment for Women Fall? An Analysis of the Incidence of the Costs of Unequal Pay and Sex Discrimination in the UK. *Gender, Work and Organisation, 2*(3), 113–124.

Burchell, B., Elliott, J., Rubery, J., & Wilkinson, F. (1994). Management and Employee Perceptions of Skill. In R. Penn, M. Rose, & J. Rubery (Eds.), *Skill and Occupational Change*. Oxford: Oxford University Press.
Clegg, S. R. (1989). *Frameworks of Power*. London: Sage Publications Ltd.
Cockburn, C. (1983). *Brothers: Male Dominance and Technological Change*. London: Pluto Press.
Cockburn, C. (1988). The Gendering of Jobs: Workplace Relations and the Reproduction of Sex Segregation. In S. Walby (Ed.), *Gender Segregation at Work*. Oxford: Oxford University Press.
Cockburn, C. (1991). *In the Way of Women*. Hampshire and London: Macmillan Press.
Cockburn, C. (1994). The Relations of Technology: What Implications for Theories of Sex and Class? In R. Crompton & M. Mann (Eds.), *Gender and Stratification*. Cambridge: Polity Press.
Colling, T., & Dickens, L. (1989). *Equality Bargaining – Why Not?* London: HMSO.
Colling, T., & Dickens, L. (2001). Gender Equality and Trade Unions: A New Basis for Mobilisation? In M. Noon & E. Ogbonna (Eds.), *Equality, Diversity and Disadvantage*. Basingstoke: Palgrave.
Collinson, M., & Collinson, D. (1996). It's Only Dick: The Sexual Harassment of Women Managers in Insurance Sales. *Work, Employment and Society, 10*(1), 29–56.
Collinson, D., Knights, D., & Collinson, M. (1990). *Managing to Discriminate*. London: Routledge.
Connell, R. W. (1983). *Which Way Is Up?* London: Allen and Unwin.
Connell, R. W. (1985). Theorising Gender. *Sociology, 19*(2), 260–272.
Coote, A., & Campbell, B. (1987). *Sweet Freedom: The Struggle for Women's Liberation*. Oxford: Blackwell Publishing.
Craig, C., Garnsey, E., & Rubery, J. (1984). *Payment Structures and Smaller Firms: Women's Employment in Segmented Labour Markets*. Research Paper No. 48. London: Department of Employment.
Crenshaw, K. (1991). Mapping the Margins: Intersectionality, Identity Politics and Violence Against Women of Color. *Stanford Law Review, 43*, 12–41.
Crompton, R. (1997). *Women & Work in Modern Britain*. Oxford: Oxford University Press.
Crompton, R., & Sanderson, K. (1990). *Gendered Jobs and Social Change*. London: Unwin Hyman.
Dahl, R. A. (1957). The Concept of Power. *Behavioural Science, 2*(3), 201–214.

Dawson, T. (2010). Out of Sight, Out of Pocket: Women's Invisibility in the British Printing Industry and Its Effect on the Gender Pay Gap. *Historical Studies in Industrial Relations, 29/30*, 61–98.

Dawson, T. (2014). Collective Bargaining and the Pay Gap in the Printing Industry. *Gender Work and Organisation, 21*(5), 381–394.

Dex, S., Lissenburgh, S., & Taylor, M. (1994). *Women and Low Pay: Identifying the Issues*. Research Discussion Series No. 9. Manchester: Equal Opportunities Commission.

Dickens, L. (1997). Gender, Race and Employment Equality in Britain: Inadequate Strategies and the Role of Industrial Relations Actors. *Industrial Relations Journal, 28*(4), 282–291.

Dickens, L. (2005). Walking the Talk? Equality and Diversity in Employment. In S. Bach (Ed.), *Managing Human Resources*. Oxford: Blackwell Publishing.

Doeringer, P., & Piore, M. (1985). *Internal Labor Markets and Manpower Analysis* (2nd ed.). Lexington, MA: DC Heath & Co.

Drake, B. (1920). *Women in Trade Unions*. London: Virago Press.

Dunlop, J. (1966). The Task of Contemporary Wage Theory. In J. Dunlop (Ed.), *The Theory of Wage Determination*. London: Macmillan.

Edwards, P. K. (2003). The Employment Relationship and the Field of Industrial Relations. In P. K. Edwards (Ed.), *Industrial Relations: Theory and Practice*. Oxford: Blackwell Publishing.

Edwards, R., Reich, M., & Gordon, D. (1975). *Labor Market Segmentation*. Lexington, MA: D Heath.

England, P. (1982). The Failure of Human Capital Theory to Explain Occupational Sex Segregation. In J. Humphries (Ed.), *Gender and Economics*. Aldershot: Edward Elgar.

Firestone, S. (1970). *The Dialectic of Sex*. London: The Women's Press.

Fosh, P., Morris, H., Martin, R., Smith, P., & Undy, R. (1983). Politics, Pragmatism and Ideology: The Wellsprings of Conservative Union Legislation 1979–1992. *Industrial Law Journal, 22*(1), 14–31.

Foster, D., & Williams, L. (2010). The Past, Present and Future of Workplace Equality Agendas: Problems of Intersectionality in Theory and Practice. In P. Blyton, E. Heery, & P. Turnbull (Eds.), *Reassessing the Employment Relationship*. Basingstoke: Palgrave Macmillan.

Foucault, M. (1994). Interview with Michel Foucault. In J. D. Faubion (Ed.), *Power: Essential Works of Foucault 1954–1984*. London: Penguin Books Ltd.

Fox, A. (1974). *Beyond Contract: Power, Work and Trust Relations*. London: Faber & Faber.

Fraser, N. (2008). From Redistribution to Recognition: Dilemmas of Justice in a "Postsocialist" Age. In K. Olson (Ed.), *Adding Insult to Injury: Nancy Fraser Debates Her Critics*. London: Verso.
French, J. P. R., Jr., & Raven, B. (1960). The Bases of Social Power. In D. Cartwright & A. Zander (Eds.), *Group Dynamics*. New York: Harper & Row.
Gallie, D. (1991). Patterns of Skill Change: Upskilling, Deskilling or the Polarization of Skills? *Work, Employment and Society, 5*(3), 319–351.
Gardiner, J. (1975). Women's Domestic Labour. *New Left Review, 89*, 47–58.
Gintis, H. (1987). The Nature of the Labor Exchange and the Theory of Capitalist Production. *Review of Radical Political Economics, 8*, 36–54.
Göhler, G. (2009). Power To and Power Over. In S. R. Clegg & M. Haugaard (Eds.), *The Sage Handbook of Power*. London: Sage Publications Ltd.
Gordon, D. (1972). *Theories of Poverty and Underemployment*. Lexington, MA: D C Heath.
Grieco, M., & Whipp, R. (1986). Women and the Workplace: Gender and Control in the Labour Process. In D. Knights & H. Willmott (Eds.), *Gender and the Labour Process*. Aldershot: Gower.
Hacker, S. L. (1990). *Doing It the Hard Way*. London: Unwin Hyman.
Hakim, C. (1979). *Occupational Segregation: A Comparative Study of the Degree and Pattern of the Differentiation Between Men and Women's Work in Britain, the United State and Other Countries*. Research Paper No. 9. London. Department of Employment.
Harris, C., & Morris, L. (1994). Households, Labour Markets and the Position of Women. In R. Crompton & M. Mann (Eds.), *Gender and Stratification*. Cambridge: Polity Press.
Hartmann, H. (1979). The Unhappy Marriage of Marxism and Feminism: Towards a More Progressive Union. *Capital and Class, 8*, 1–33.
Hartmann, H. (1981). The Unhappy Marriage of Marxism and Feminism: Towards a More Progressive Union. In L. Sargent (Ed.), *Women and Revolution: A Discussion of the Unhappy Marriage of Marxism and Feminism*. Boston: South End Press.
Haugaard, M. (2009). Power and Hegemony. In S. R. Clegg & M. Haugaard (Eds.), *The Sage Handbook of Power*. London: Sage Publications Ltd.
Haugaard, M., & Clegg, S. R. (2009). Why Power Is the Central Concept of the Social Sciences. In S. R. Clegg & M. Haugaard (Eds.), *The Sage Handbook of Power*. London: Sage Publications Ltd.
Heery, E. (2006). Equality Bargaining: Where, Who, Why? *Gender, Work and Organization, 13*(6), 522–542.

Heery, E., & Kelly, J. (1989). Full-Time Officers and Trade Union Recruitment. *British Journal of Industrial Relations, XXVII*(2), 196–213.

Holgate, J., Hebson, G., & McBride, A. (2006). Why Gender and 'Difference' Matters: A Critical Appraisal of Industrial Relations Research. *Industrial Relations Journal, 37*(4), 310–328.

Hyman, R. (1975). *Industrial Relations: A Marxist Introduction*. London: Macmillan Press.

Hyman, R. (1994). Changing Trade Union Identities and Strategies. In R. Hyman & A. Ferner (Eds.), *New Frontiers in European Industrial Relations*. Oxford: Blackwell.

Jenkins, R. (2009). The Ways and Means of Power: Efficacy and Resources. In S. R. Clegg & M. Haugaard (Eds.), *The Sage Handbook of Power*. London: Sage Publications Ltd.

Jewson, N., & Mason, D. (1992, originally 1986). The Theory and Practice of Equal Opportunities Policies. In P. Braham, A. Rattansi, & R. Skellington (Eds.), *Racism and Antiracism*. Milton Keynes: Open University Press.

Karamessini, M. (2014). Introduction – Women's Vulnerability to Recession and Austerity: A Different Crisis, a Different Context. In M. Karamessini & J. Rubery (Eds.), *Women and Austerity: The Economic Crisis and the Future for Gender Equality*. Oxford: Routledge.

Keep, E. (2005). Skills, Training and the Quest for the Holy Grail of Influence and Status. In S. Bach (Ed.), *Managing Human Resources*. Oxford: Blackwell.

Keep, E., & Rainbird, H. (1995). Training. In P. Edwards (Ed.), *Industrial Relations Theory and Practice in Britain*. Oxford: Blackwell.

Kelly, J. (1998). *Rethinking Industrial Relations: Mobilization, Collectivism and Long Waves*. London: Routledge.

Kimmel, M. S., & Coston, B. M. (2017). Seeing Privilege Where It Isn't: Marginalized Masculinities and the Intersectionality of Privilege. In M. S. Kimmel & A. L. Ferber (Eds.), *Privilege: A Reader*. Boulder, CO: Westview Press.

Kirton, G. (1999). Sustaining and Developing Women's Trade Union Activism: A Gendered Project? *Gender, Work and Organization, 6*(4), 213–223.

Kirton, G., & Greene, A. (2016). *The Dynamics of Managing Diversity* (4th ed.). London: Routledge.

Kirton, G., & Healy, G. (1999). Transforming Union Women: The Role of Women Union Officials in Union Renewal. *Industrial Relations Journal, 30*(1), 31–45.

Ledwith, S. (1991). *Sweet Talk, Soft Deals? – Gender and Workplace Industrial Relations*. Working Papers in Economics, Business and Management No. 35. London: Polytechnic of North London.

Ledwith, S., & Colgan, F. (2002). Tackling Gender, Diversity and Trade Union Democracy. In F. Colgan & S. Ledwith (Eds.), *Gender, Diversity and Trade Unions*. London: Routledge.

Loveridge, R., & Mok, A. L. (1979). *Theories of Labour Market Segmentation*. London: Kluwer Academic.

Lown, J. (1983). Not So Much a Factory, More a Form of Patriarchy: Gender and Class During Industrialisation. In E. Gamarnikow et al. (Eds.), *Gender, Class and Work*. London: Heinemann Educational Publishers.

Lukes, S. (2005). *Power: A Radical View*. London: Palgrave Macmillan.

Martin, R. (1992). *Bargaining Power*. Oxford: Oxford University Press.

McCall, L. (2005). The Complexity of Intersectionality. *Signs, 30*(3), 1771–1800.

McEwen Scott, A. (1994). Industrialization, Gender Segregation and Stratification Theory. In R. Crompton & M. Mann (Eds.), *Gender and Stratification*. Cambridge: Polity Press.

Millet, K. (1971). *Sexual Politics*. London: Virago.

Mincer, J., & Polachek, S. (1980, originally 1974). Family Investments in Human Capital: Earnings of Women. In A. H. Amsden (Ed.), *The Economics of Women and Work*. Middx: Penguin.

Munro, A. (1999). *Women, Work & Trade Unions*. London: Mansell.

Noon, M. (2007). The Fatal Flaws of Diversity and the Business Case for Ethnic Minorities. *Work, Employment and Society, 21*(4), 773–784.

Office for National Statistics (ONS). (2013, 2014). Annual Survey of Hours and Earnings by Industry – Tables 16.1a Gross Weekly Earnings, Not Affected by Absence and 16.5a Gross Hourly Earnings, Not Affected by Absence.

Parker, J. (2009). Women's Collectivism in Context: Women's Groups in UK Trade Unions. *Industrial Relations Journal, 40*(1), 78–97.

Phelps Brown, H. (1977). *The Inequality of Pay*. Oxford: Oxford University Press.

Phillips, J. (2014). Containing, Isolating, and Defeating the Miners: The UK Cabinet Ministerial Group on Coal and the Three Phases of the 1984–1985 Strike. *Historical Studies in Industrial Relations, 35*, 117–141.

Phillips, A., & Taylor, B. (1980). Sex and Skill: Notes Towards a Feminist Economics. *Feminist Review, 6*, 79–88.

Pollert, A. (1983). Women, Gender Relations and Wage Labour. In E. Gamarnikow et al. (Eds.), *Gender, Class and Work*. London: Heinemann Education Publications.

Pollert, A. (1996). Gender and Class Revisited; or, The Poverty of Patriarchy. *Sociology, 30*(4), 639–659.

Printing and Publishing Industry Training Board (PPITB). (1982). *Equalising Job Opportunities for Women in Printing and Publishing: An Attitude Survey Report.* London: Printing and Publishing Industry Training Board.

Purcell, K. (1979). Militancy and Acquiescence Amongst Women Workers. In S. Burman (Ed.), *Fit Work for Women.* London: Croom Helm.

Rowbotham, S. (1973). *Hidden from History.* London: Pluto Press.

Rowbotham, S. (1982). The Trouble with 'Patriarchy. In M. Evanston (Ed.), *The Woman Question.* London: Fontana.

Rubery, J. (1994). Internal and External Labour Markets: Towards an Integrated Analysis. In J. Rubery & F. Wilkinson (Eds.), *Employer Strategy and the Labour Market.* Oxford: Oxford University Press.

Rubery, J. (2014). From 'Women and Recession' to 'Women and Austerity': A Framework for Analysis. In M. Karamessini & J. Rubery (Eds.), *Women and Austerity: The Economic Crisis and the Future for Gender Equality.* Oxford: Routledge.

Rubery, J., & Fagan, C. (1995). Comparative Industrial Relations Research: Towards Reversing the Gender Bias. *British Journal of Industrial Relations, 33*(9), 209–236.

Rubery, J., & Rafferty, A. (2013). Women and Recession Revisited. *Work, Employment and Society, 27*(3), 414–432.

Sawyer, M. (1989). *The Challenge of Radical Political Economy.* Hemel Hempstead: Harvester Wheatsheaf.

Seccombe, W. (1974). The Housewife and Her Labour Under Capitalism. *New Left Review, 83*, 3–24.

Smith, P. (1978). Domestic Labour and Marx's Theory of Value. In A. Kuhn & A. Wolpe (Eds.), *Feminism and Materialism.* London: Routledge and Kegan Paul.

Smith, P., & Morton, G. (2006). Nine Years of New Labour: Neoliberalism and Workers' Rights. *British Journal of Industrial Relations, 44*(3), 401–420.

Speakman, S., & Marchington, M. (1999). Ambivalent Patriarchs: Shiftworkers, 'Breadwinners' and Housework. *Work, Employment and Society, 13*(1), 83–105.

Stanko, E. A. (1988). Keeping Women In and Out of Line: Sexual Harassment and Occupational Segregation. In S. Walby (Ed.), *Gender Segregation at Work.* Milton Keynes: Open University Press.

Storey, J. (1983). *Managerial Prerogative and the Question of Control.* London: Routledge and Kegan Paul.

Thompson, P. (2011). The Trouble with HRM. *Human Resource Management Journal, 21*(4), 355–367.

Thornley, C. (1996). Segmentation and Inequality in the Nursing Workforce: Re-Evaluating the Evaluation of Skills. In R. Crompton, D. Gallie, & K. Purcell (Eds.), *Changing Forms of Employment: Organisations, Skills and Gender*. London: Routledge.

Thornley, C. (2002). *Double Jeopardy: Non-Registered Nurses in the NHS.* Conference Paper, Gender Research Forum, London.

Thornley, C. (2006). Unequal and Low Pay in the Public Sector. *Industrial Relations Journal, 37*(4), 344–358.

Trade Union Act. 2016. http://www.legislation.gov.uk/ukpga/2016/15/contents/enacted.

Trades Union Congress. (2016). *Equality Audit.* London: Trades Union Congress.

Unison. (2017). Retrieved November 15, 2017, from https://www.unison.org.uk/news/article/2017/08/employment-tribunal-fees-story-behind-success/.

Wajcman, J. (2000). Feminism Facing Industrial Relations in Britain. *British Journal of Industrial Relations, 38*(2), 183–201.

Walby, S. (1986). *Patriarchy at Work.* London: Polity.

Walby, S. (1988). *Gender Segregation at Work.* Milton Keynes: Open University Press.

Walby, S. (1990). *Theorizing Patriarchy.* Oxford: Blackwell.

Walton, R., & McKersie, R. (1965). *A Behavioral Theory of Labor Negotiations.* New York: McGraw Hill.

Willis, P. (1977). *Learning to Labour.* Farnborough: Saxon House.

Windebank, J. (2001). Dual-Earner Couples in Britain and France: Gender Divisions of Domestic Labour and Parenting Work in Different Welfare States. *Work, Employment and Society, 15*(2), 269–290.

Witz, A. (1992). *Professions and Patriarchy.* London: Routledge.

Wootton, B. (1955). *The Social Foundations of Wage Policy.* London: George Allen & Unwin.

Young, I. (1981). Beyond the Unhappy Marriage: A Critique of the Dual Systems Theory. In L. Sargent (Ed.), *Women and Revolution: A Discussion of the Unhappy Marriage of Marxism and Feminism.* Boston, MA: South End Press.

3

The Development of the Printing Industry: Workers' and Employers' Organisation

Printing has a history of over 500 years in Britain. Tradition has always been influential in the industry, and work organisation has been broadly stable from the late nineteenth century to the turn of the millennium. The practices associated with the labour process established the territory on which women would enter the industry.

What follows is a discussion of the nature of the labour process and its changes, how men entered it, how they organised themselves and how employers sought to control the industry over time. This will primarily focus on the establishment of class relations and will link closely to Chap. 4, which will identify how women came to the industry, how they were utilised by employers and how they sought to improve their conditions of work. What becomes apparent is that male workers in printing were primarily concerned to protect their positions at any cost within a class dynamic that exposed their vulnerability in a capitalist economy. This is the context within which male workers acquired their access to power resources, in a struggle against employers' economic power and its attendant power resources. Women were casualties and seen as expendable in this crucial battle. The contradictions of the sectionality that lay at the heart of this approach are explored in Chap. 4.

T. Dawson, *Gender, Class and Power*, https://doi.org/10.1057/978-1-137-58594-3_3

First it is important to grasp the special position printing occupied in its earliest days. William Caxton's introduction of the printing press to England in the fifteenth century helped fuel the political and religious upheavals of the sixteenth and seventeenth centuries. The arrival of the printing press was not just another technical innovation but carried implications of much greater import. The character of the industry was established during this period, creating the context for women's entry into the industry.

Early Development: 1470–1780

Before industrialisation books were either produced in monasteries or as a bespoke item. There were three parts to this manufacturing process that have remained recognisably intact ever since—composition, presswork and bookbinding.[1] While the first two were brought together with the establishment of mechanical printing in Britain, bookbinding often remained separate as it sometimes was before industrialisation (Child 1967; Delafons 1965). There was, therefore, a seemingly 'natural' division of labour from the outset. Women were only ever present in substantial numbers in bookbinding (Cockburn 1983).

William Caxton established the first press in England in 1476 in Westminster (Steinberg 1996), while Edinburgh saw the establishment of the first press in Scotland in 1507 (Gillespie 1953). By the 1550s more printing presses had been introduced, initially to the universities at Oxford and Cambridge, but also to other 'provincial towns, usually in the vicinity of ecclesiastical buildings' (Musson 1954: 4).

By 1557 the Stationers' Company, one of the craft guilds,[2] had been formed, and printing was strictly regulated with the Company having almost a complete monopoly of the practice of 'the art and mystery of printing' (ibid.). Printing could only be carried out under licence, and any material that was in breach of the law could be confiscated and destroyed. From 1586 the Star Chamber curtailed provincial printing, apart from at Oxford and Cambridge. Even in London printing was further restricted. This must be seen in the context of state censorship, which arose from the fear of spreading dissension among the people during the religious upheavals of the period. In particular, Henry VIII needed to control the means of propaganda in his battles with the Catholic Church in Rome (Child 1967; Hibbert 1992). However, printers did continue to set up and publish 'political and religious works of a seditious character' (Musson 1954: 5).

The need to ensure full employment to prevent printers' engagement in illegal publishing gave rise to the first restrictive practices around apprenticeship. It also preserved the Stationers' Company's monopoly. For instance, apprentices were not to be employed if any journeyman[3] lacked employment (ibid.: 8). In 1587 the Stationers' Company actively encouraged restrictive practices as 'the journeyman's right to employment took priority over the commercial interest of the master' (Child 1967: 26).

During this period the seven years' apprenticeship became commonplace, but later, as industrialisation developed, employers often quite successfully employed boys with only two or three years' experience. This also suited the boys because their wages rose faster (Musson 1954), but meant they also had potential as a reserve army (Cockburn 1983). The length of the apprenticeship therefore needs to be seen in the light of a desire for control, rather than as an indicator of the level of skill required for the job and also as a form of cheap labour for employers (Phillips and Taylor 1980; Delafons 1965).

The 1688 Glorious Revolution finally destroyed state censorship allowing an expansion of printing. There was, however, a natural check on its growth and that was the low level of literacy within the population. '[A]lthough, already impressive by European standards by 1700, British literacy levels did not rise very sharply in the next century' (Porter 2000: 76). However, demand for a greater variety of reading material increased. As a result a number of newspapers began publishing in the eighteenth century. The technology developed during this period remained remarkably stable right through to the end of the eighteenth century for presses and to the end of the nineteenth century for composing equipment (Musson 1954).

State control, through the guilds, had been extensive comprising regulation of all aspects of printing including working conditions and wages. The most important figure around whom this system revolved was the master printer who was responsible for everything from finding work to planning and executing it (Eisenstein 1983).

All had their role in this system. Employers (masters) and workers (journeymen and apprentices) had all been part of the Stationers' Company 'which although autocratic was also paternal … [aiming] to keep a steady flow of work for the journeymen and to provide for sickness and old age' (Moran 1964: 4). Then the guild system broke down in the late seventeenth century. The gradual demise of the Stationers' Company left journeymen feeling vulnerable. There was a breakdown of traditional customs, and the inflation caused by the French wars (Child 1967) gave rise to calls for higher wages. The combinations of workers, forerunners to the trade unions, formed at this time, had some initial success and so became established. Many of the guild's rules were subsequently adopted by these trade societies.[4] As Child recognises:

> The roots of the beliefs and traditions of the early craft societies lie deep in history, in the ideal of the stable social framework, where each man had a right to his trade, where prices and wages were fixed by law, where the supply of labour was regulated to suit demand. (ibid.: 60)

Before the guilds demise, and certainly by 1682, the particular form of workplace organisation, called the chapel, was in place, although the idea of it being used for trade protection had yet to develop. Chapels were a form of workplace union branch, generally consisting of a group of workers undertaking the same or similar work in a contained part of the labour process. Instead of industrial action, printers would resort 'to a series of petty embarrassments' (Child 1967: 26) rather than strike, and, if that failed, they would have recourse to law. It was only later that chapels gradually 'came to regard the protection of their members as their principal, though not sole, object' (Avis 1971: 22).

Whatever the derivation of 'chapel', certainly the rituals attached to initiating one and for admitting members had a religious flavour and resembled the Freemason's elaborate customs. Initiation ceremonies for apprentices that were a travesty of religious rites were required, and where a press was to be set up in a house not previously used for printing, the room had to be consecrated by another ceremony (Musson 1954; Child 1967). Taken from *The Gentleman's Magazine*, Musson quotes the following:

> When a Printer first sets up, if it is an House that was never used for Printing before, the Part designed for that Purpose is consecrated, which is performed by the Senior Freeman the Master employs, who is the Father or Dean of the Chapel; and the chief Ceremony is drinking Success to the Master, sprinkling the Walls with strong Beer, and singing the Cuz's Anthem. [There follows an account of the Chapel 'Laws', and penalties imposed for their breach] … which an obstinate Member sometimes refuses to pay; upon which it is left to the Majority of the Chapel, in Convocation assembled, whether he shall be continued any longer a Chappellonian; and if his Sentence is to be discontinued, he is then declared a Brimstone; that is, an excommunicated Person. (pp. 13–14)

And in order to become a member of the chapel there were ceremonial rites to perform:

> When a Boy is to be bound Apprentice, before he is admitted a Chappellonian, it is necessary for him to be made a Cuz, or Deacon; in the Performance of which there are a great many Ceremonies. The Chappellonians walk three Times round the Room, their right arms being put thro' the Lappets of their Coats; the Boy who is to be made a Cuz carrying a wooden Sword before Them. Then the Boy kneels, and the Father of the Chapel, after exhorting him to be observant of his Business, and not to betray the Secrets of the Workmen, squeezes a Spunge of strong Beer over his Head, and gives him a Title, which is generally that of Duke of some Place of the least Reputation near which he lives, or did live before … Whilst the Boy is upon his Knees, all the Chappellonians, with their right Arms put through the Lappets of their Coats, as before, walk round him, singing the Cuz's Anthem, which is done by adding all the Vowels to the Consonants in the following Manner.
>
> Ba—ba; Be—be; Bi—bi; Ba—be—bi; Bo—bo; Ba—be—bi—bo; Bu—bu; Ba—be—bi—bo—bu—And so through the rest of the Consonants. (p. 14)

Beer featured strongly in all this ceremonial, with chapel funds largely spent on its consumption, and it likely helped cement the chapel and its customs (Musson 1954). In fact, so strong was the feeling for this tradition that when the London lodge of the bookbinders' union tried to move away from the local pub as meeting place, it brought division between the members for ten years (Child 1967). Lodges were the bookbinders' equivalent of chapels and had similar elaborate rituals (ibid.).

The members democratically controlled the chapel. There were a number of basic rules that seemed to apply generally and covered fighting, swearing and lying, as well as shoddy work practices (ibid.: 37). Many of these rules go back so far that even Joseph Moxon, whose seventeenth-century publication *Mechanick Exercises* was the authority on the printing industry, could not trace their origins (ibid.: 35). Breach of these laws was punished by a solace (penalty) determined by the whole chapel membership. All journeymen had to join the chapel and obey the rules, of which there were many. The chapel ghost, Ralph, would keep in line anyone who refused to join or broke the rules. He was responsible for hiding the workman's tools or mixing up his type among 'other annoying tricks' (ibid.: 36). The recalcitrant journeyman was treated like an excommunicate, in keeping with the spurious religious origins of the chapel. Invariably the wayward son would be brought into the fold after a few weeks in Ralph's ethereal company.

The key figure in the running of the chapel was the father of the chapel (FoC), so-called, as initially, this role went to 'the oldest freeman-printer' (Avis 1971: 29) in the workplace. In most industries this would equate to the shop steward. This was in keeping with the paternalism of the age, and the position did not become open to annual election for many years. The FoC was 'variously arbiter, counsellor, examining and sentencing magistrate, spokesman, or mediator, depending on the nature of the cause' (ibid.: 30) in any dispute, as well as calling and chairing meetings, liaising with management, providing advice to members and circulating information. The role of the FoC has remained fairly constant right through to the 1980s, since when the role has diminished. Decisions were by democratic vote.

Apprentices were not chapel members although the chapel monitored the ratio of apprentices to journeymen (Avis 1971). On signing indentures there was an initiation ceremony, and at the end of their time, they were 'banged out', a ceremony that involved a lot of noise and could continue for some time (ibid.: 37). These ceremonies related to the trade and not to chapel membership. In addition, the apprentice lived in the master's house (Duffy 2000).

Laissez-faire economics became the new orthodoxy at this time. Industry grew rapidly, employers had a free hand but not so the workers. It is likely that some form of journeymen's organisation began to form during the second half of the eighteenth century as the guild collapsed, particularly given the growth of provincial printing. No evidence of trade unionism among printers or bookbinders can be found until around 1785, and given the subsequent legal restrictions of the Combination Acts[5] (Musson 1954), lack of documentary evidence is not surprising. Nevertheless, all of the old 'traditions survived the impact of *laissez-faire* and the industrial upheaval of the nineteenth century' (Child 1967: 60), and printers still strove for stable social relations that recognised their status and claim to the trade.

Technological change also presented challenges. The move from hand to mechanical presses in the late eighteenth century led to the introduction of a new type of skilled worker—the machine minder (Musson 1954), although this was most advanced in London newspapers (Moran 1964). This segmentation resulted in machine minders gradually becoming a category of worker in their own right. Initially compositors objected to this specialisation, insisting that printing could only really be done by time-served men who understood the whole process—minders were nothing more than crude mechanics (Musson 1954). But at the same time they disdained machine work so that non-time-served men gained access to these jobs, a problem that would return to haunt compositors during the nineteenth century.

The small size of workshops meant it was possible for a thrifty journeyman to aspire to become a master (ibid.: 23). The average print shop would consist of 'five compositors, two pressmen and three apprentices' (Child 1967: 48). A plethora of small companies led the expansion of the printing industry, but this hindered the growth of unions, as larger offices were easier to organise, often reducing the scope of industrial relations to petty disputes with little consistency of outcome. The industry's expansion resulted in the 'gradual emergence of a body of journeymen having small prospect of eventual ownership of a business [which] strained existing ideas of trade relations' (Avis 1971: 22) and began to alter the role of the chapel.

This account of developing industrial relations in the early period of printing demonstrates the development of an overwhelmingly masculine culture. The drinking, initiation ceremonies and policing of behaviour would all be likely to deter women from even considering such a job. Crucially, it also identifies the centrality of technical power. First the Stationers' Company carefully controlled entry to the trade and ability to gain employment thereafter but also established the roots of subsequent protectionism through restrictive practices. This is what provided, and then cemented, men's access to technical power and, latterly, allowed them to outflank women. As the guild collapsed, these practices became the bedrock of the embryonic trade unionism that eventually replaced it. The disappearing possibility of becoming a master highlighted the need for collective protection, and the trade societies became vehicles for the exercise of sexual and collective power resources. Segmentation in printing, between compositors and machine minders, also reinforced and justified these processes. This would become a decisive factor in gender relations as capitalism developed.

The Nineteenth Century

Frequent trade depressions throughout the nineteenth century caused immense hardship and led to the development of the tramping system, whereby men would receive a small sum of money at each union-organised town if they failed to find work. They would then be expected to move on. It was an inadequate strategy (Musson 1954), but the aim was to 'prevent … [workers] accepting employment, under stress of starvation, on terms which … would be injurious to … [the union's] interests' (Webbs 1913: 161). The fear of unemployment was palpable as unions railed against 'rats' who broke strikes, 'forreigners' (sic) coming into the industry and 'turnovers' (apprentices who changed master before finishing their time), such restriction being the 'pivot of … Trade Unionism' (ibid.: 453) at this time. At the same time '[t] he craftsmen believed, with religious intensity and unquestioning faith, that only the servitude of a formal apprenticeship conferred on a man the precious "right to the trade"' (Child 1967: 66).

Consequently, there arose rules prohibiting women, boys and unskilled men from undertaking skilled work, as well as turning the union into an agency to supply labour through use of the Call Book[6] (ibid.: 41, 74), a system continued well into the twentieth century. Meanwhile the employers undermined strike action through the use of the unemployed, blacklisting, casualisation and non-time-served labour (ibid.: 69–70; Musson 1954). One tactic advanced by the trade societies at this time was to pay strikebreakers not to undertake their work during a dispute (Turner 1962), and the printers were no exception. Unfortunately, the low level of strike pay was an insufficient response. Job insecurity drove a wedge between these different groups of workers, encouraging identification of the craftsman's 'enemies' among 'the less advantaged classes of workers' (Gordon 1972: 80). Consequently, while printers were a self-defined elite and mostly better paid than other industrial workers, most of their actions were defensive, aimed at maintaining wage rises and preserving privileges (Child 1967).

Throughout the nineteenth century unemployment undermined control of apprentices, overtime and wages and necessitated printers' organisation. An early attempt to go beyond simple local organisation occurred in the 1830s with the formation of the Northern Typographical Union, but this was undermined by economic depression between 1836 and 1843 and lack of expertise in finance and central administration. Finally, the need for central control of the tramping system created pressure to amalgamate, and eventually the Provincial Typographical Association (PTA) emerged in 1849 comprising 'only eight independent Societies ... [with] a total membership of but four hundred and eighty one' (Sessions 1950: 32). However, local autonomy still tended to undermine concerted action (Musson 1954). Similarly, the Bookbinders Consolidated Union was a craft union, fully established by 1855, initially concerned with operating tramping funds and reluctantly drawn to industrial protectionism (Bundock 1958). Tramping started to die out during the nineteenth century (Musson 1954; Bundock 1958). Politically, the TA, for instance, was repulsed by anything that smacked of revolution or violence. They also deplored coercion of employers and preferred co-operation. How much this was influenced by inherent union weakness, rather than principle, is not clear. Instead, 'Industry, sobriety and self-education—these were to be the means of liberating workers from poverty and vice' (Child 1967: 89).

During the nineteenth century certain features of craft trade unionism became apparent. Numerous rules that have shaped the form of later national agreements appeared at this time, for example, the system of extra payments and balancing time[7] (Musson 1954). The unions deplored piece rates but were unable to stamp them out. It was to avoid exploitation of piece workers that the complicated process of 'extra' payments developed (ibid.: 210–217). Such concessions were probably extracted only to secure crucial skills (Pelling 1976). Similarly, the PTA tried to enforce apprentice restrictions but failed, particularly when technology changed and the industry expanded. For instance, many of the early machine minder members had not served an apprenticeship but had obtained positions by 'progression'[8] (Musson 1954).

Early TA rules demonstrate the printers' attitude to membership eligibility: a rather extreme form of closed unionism. As well as serving a seven-year apprenticeship, the applicant had to be earning the 'minimum standard of wages and working in 'fair' offices' (Musson 1954: 116). To which one Nottingham employer remarked, 'it appeared rather a novel proceeding that the firm should be asked to raise the wages in order to enable the men to join the branch' (ibid.: 117), rather than vice versa. This attitude had to be relaxed to allow for expansion of membership. Despite the technological change of the 1890s, members were wedded to their work as a craft, not a trade.

This is the period in which access to primary power resources became embedded with stronger trade union organisation. The prime concern of print unions was to keep control of the labour supply as the labour process developed. Despite the craft unions' attempts at such control, it was clearly an ongoing battle against the cheapening of labour, and questions as to appropriateness of their approach are discussed in Chap. 4. Segmentation started to make inroads into union organisation, too. In essence, fluctuating economic performance created swings between skill shortages and oversupply of labour, and so trade union development was uneven. However, stronger unionism would start to extract greater concessions from the late nineteenth century onwards.

The Early Twentieth Century

The first development was a new employers' organisation. Generally employers' associations were ad hoc, local affairs until the late nineteenth century. In printing, only London had any form of employer organisation before the nineteenth century (Clegg 1976). Print employers were trying to develop organisations, with little unity prior to this. The masters would 'come together intermittently, when a crisis threatened; at other times the state of the labour market was sufficiently in their favour to make further organization unnecessary' (Child 1967: 48).

Also union demands were viewed as reasonable and helped avoid undercutting, at least until about 1890 when the TA 'achieved its full scope in negotiation' (Sessions 1950: 33). Crucially, '1897 was a year of labour unrest, when demands were being made for the increase of wages and the reduction of hours' (ibid.: 4), ultimately forcing the formation of a national federation in 1901 (Howe 1950). By 1911 there was also a sister organisation, the Scotland Alliance of Employers (Gillespie 1953).

A leading master printer, Walter Hazell, made the following statement of his hopes for the new federation:

> We, as a confederation, can do what we are unable to do individually. If it had been our purpose to make ourselves a monopoly against the interests of the public or if we had aimed to injure the Trade Unions, I personally would have had nothing to do with the Federation. These, however, are not our objects. I think that by the Federation we shall be able to do a great deal of good to the community of which we form a part. No class is so put upon as the printers, and I hope that it will be possible to formulate some set of rules as to what is and what is not reasonable in regard to the demands of our more unreasonable customers. I believe in Trade Unions, but if they obtain too much power, they are likely to abuse it, and it is essential that there should be associations of employers to meet the employees on equal terms. We all desire to improve the lot of our work-people, and that all concerned shall do their work so well that they can hold their own, give better wages, better conditions, and make life more tolerable to their work-people and more satisfactory to themselves. (Sessions 1950: 12–13)

More prosaically, the need to deal with technological change, circulate information and mirror employer organisation elsewhere were key reasons for federation (Howe 1950). Primarily employers' federations formed at around this time to confront the growth in trade unionism (Clegg 1976), and Sessions' (1950) records show that of 70 local associations for which she had information, 50 started expressly to deal with labour issues.

From the start of employer organisation, there were some types of national agreement being reached with the unions (Howe 1950). However, the paternalism inherent in Walter Hazell's early address was under strain by the time of the president's statement in 1916 that 'we must strengthen our defences against the increasing, and often unreasonable demands of those we employ' (Sessions 1950: 19). By 1911, after government intervention on the issue, the first national agreement appeared on hours (Musson 1954). It was not easy to reach agreement. For instance, an early opposition of the Federation of Master Printers to reducing hours was justified on the grounds that the P&KTF[9] 'did not represent any strong proportion of trade-union labour throughout the country, and that many shops only employed about 15 per cent of union hands' (Howe 1950: 23).

Meanwhile there was a widespread problem of undercutting (ibid.: 7–8). By 1913 the Federation was convincing its members to introduce scientific costing[10] but had to reassure the unions that it was not just another way to speed up the machines. Reassurance may have come from the industry's place in the wages league, as Walter Hazell pointed out, the costing system had ensured that printing 'gave an adequate return for the capital invested in it ... and the wages of its craftsmen compare favourably with those of any other industry' (Sessions 1950: 51).

No trade union had a proper full-time administration until the twentieth century (Bundock 1958), and organising was not something considered all that necessary initially. This was apparent in the TA in 1891 when they appointed two organisers, only to backtrack in 1895 because they were too expensive. They realised their mistake by 1898, employing two in England and Wales and one in Ireland, assisting rapid continuous membership growth through to 1914. It was claimed that they had around 66 per cent membership density by 1904. But when it came to subscriptions members were reluctant to raise them, despite the need to support benefit payments, so the union lurched from one financial crisis to another. Meanwhile, in terms of industrial matters, there was a constant battle of output restriction and 'sweating'. Organising in machine rooms also continued to be a concern, especially in relation to 'manning', and demarcation. The latter was resolved with an agreement between the National Society of Operative Printers and Media Personnel (NATSOPA) and the TA, much to the employers' annoyance.

The other major development came at the end of 1918—national bargaining. In common with much of industry (Clegg 1976), negotiations had been done by the local trade union societies, often starting with the big cities, while the smaller towns were left to fight it out for themselves (Howe 1950). Mirroring developments elsewhere, the employers were the main movers in the strategy of national bargaining, aiming to restrict local militancy. The printers came reluctantly to conclude that it would be best to move to a national wage agreement, the first of which was signed in 1919 (ibid.). The Federation's new labour secretary, a former printers' union executive council member, drew up a grading system (ibid.). It graded by town size and established a machine classification agreement that continued the tradition of extra payments, now related to the characteristics of the machine operated.

The 1914–1918 war caused disruption but is discussed in more detail in the next chapter because of the important involvement of women. The interwar years were characterised by attempts to fight off pay cuts and achieve reduced hours (Musson 1954). This was fuelled by continually improving technology and gave rise to employer arguments that the work was easier, countered by the claim that printers should share in the increased profits. Although the early part of this period was characterised by a series of wage increases, followed by a series of wage cuts, from 1923 to 1940 wages remained stable, probably assisted by union density of between 71 per cent and 88 per cent. However, wage cuts had disturbed differentials between the unions, and this would resurface as a bitter problem after the Second World War. But stable wages did not stop constant negotiations for extras on all new or complicated machinery such that '[i]t would be difficult to specify even a tithe of these negotiations' (Howe 1950: 112), probably widening differentials.

During the first half of the twentieth century, unions became embedded in British industry. Both employers and male workers increased their power resources; the men, in particular, sought to embed their technical power. Secondary power resources (personal, positional and symbolic) became more important in intra-organisational relations, as well as between the parties to negotiations. But things were about to get tougher after the Second World War, as employers sought to overcome union advantage and gradually any notions of consensus were hollowed out.

Post Second World War

Gennard (1990) suggests that the factors that had the greatest impact on the printing industry up to 1990 were the demise of letterpress[11] and the accompanying technological changes, the growth of white-collar employment, the merger of the print unions, import penetration, the growth of inplant and instant printing,[12] competition from unorganised advertising agencies and studios and the advent of mass electronic communications. But the impact of political and legal change also requires analysis.

The Gennard (1990) and Gennard and Bain (1995) histories cover the formation of both the National Graphical Association (NGA) and the Society of Graphical and Allied Trades (SOGAT) to their merger to form the GPMU. This was the ultimate aim—one union for the industry. Getting there was a difficult journey and illustrates the conservative attitudes that not only prevented amalgamation for so long but also restricted women's participation in both the industry and the union (see Chap. 4). Gennard and Hayward (2008) cover the GPMU years but fail to adequately engage with women's role in the union or industry during that period, so have little to offer the analysis in Chap. 5.

In general printing[13] in the 1950s, there was still clear demarcation of jobs. Craftworkers, from purely craft unions, did typesetting and controlled the printing process. The bookbinders' union would generally provide the assistants for both the typesetting and printing processes. The finishing (semi- and unskilled) jobs such as 'folding, counting, collating, gathering, stitching, binding, trimming, packing and despatch … [were mostly] done by hand by semi-skilled or unskilled workers, most of whom were women' (Gennard and Bain 1995: 5). In 1950 '[n]early 70 per cent of firms in membership of the British Master Printers' Federation[14] had fewer than 25 employees', and '[l]ess than 3 per cent employed 300 or more' (Gennard 1990: 3–4). Employment in the printing industry at this time was approximately 375,000 with over 243,000 in general printing. The rest of the unionised workforce was split among at least ten craft unions, reduced to five by 1969 through mergers. So what drove the impetus to merge?

Post-1945 some unions 'did not want to participate in a collective wages movement and by October 1947 the employers faced 13 uncoordinated wage claims' (Forster 1956: 5). The employers knew that granting any one of them would lead to a series of 'leapfrogging claims' and by the mid-1950s would only concede the same deal to each union. The employers' federations[15] had twice attempted to propose industry-wide wage structures, but they were always unacceptable to the unions. The struggle towards a workable wage structure was the preoccupation of both unions and employers throughout the 1950s, but it was undermined by narrow sectional interests, especially the craft unions' fear of domination of the bargaining agenda by the low paid (Forster 1956). Despite the benefits of 'leapfrogging', the unions gradually realised that the:

> ability of one union to improve, and sustain, the relative advantage in terms of its members' wages and conditions in isolation from other print unions … [was] limited. (Gennard and Bain 1995: 63)

Nevertheless, suspicion was still the watchword between craft and non-craft unions and London and provincial unions especially. But technological change brought redundancies, and division prevented London printers from following the job into the provinces, while the non-craft union members were still prevented from training up for craft jobs (Gennard 1990; Gennard and Bain 1995).

Additionally, high relative production costs and lower quality *inter alia* in domestic production saw import penetration increase in the post-war period, by 400 per cent from 1958 to 1968 (NEDO 1970), which was still an issue in the 1980s (Craig et al. 1984). Labour shortages compounded these problems, encouraging employers to invest in new techniques and unions to concede, however reluctantly, increased labour intakes (Gennard 1990).

The result was a merger between the London and provincial printers' craft unions in 1964 to form the National Graphical Association (NGA), followed by a series of other craft mergers into the NGA (Gennard 1990). This pre-empted the non-craft union merger in 1965 between the bookbinders (NUPB&PW) and the machine assistants (NATSOPA) to form SOGAT. The latter was an ill-thought-out venture that foundered in 1972 with the continuing failure to agree union structures and rules (Gennard and Bain 1995). During this period there had been an alternative possibility of a merger between the non-craft NUPB&PW and the craft union, the Amalgamated Society of Lithographic Printers (ASLP), that Gennard suggests could have had far-reaching consequences for the early development of one union for the industry. However, this failed, leaving craft and non-craft unions in separate camps for another 25 years (Gennard 1990: 75).

Meanwhile attitudes to progression from non-craft to craft jobs, with the development of web presses,[16] caused friction and demarcation disputes between the respective unions, over access to the new equipment (Cameron 1967). Neither of the two craft unions (NGA and ASLP) was prepared to countenance equal access for non-craft unions (Gennard and Bain 1995). Contrary to the Webbs' assertion that the use of a system of progression was a positive development designed to promote continuity of employment (1913), in printing it continued to deny non-craftworkers' access to skilled jobs, preserving the privileges of craft printers until the 2000s, craftworkers claiming a moral right to the trade (ibid.) that became progressively untenable during the late twentieth century.

After the dissolution of the bookbinders' and assistants' merger in 1972, the bookbinders, formerly NUPB&PW, retained the name SOGAT and went on to merge with the Scottish printers' union, whose earlier proposed merger with the TA was scuppered because the former organised women (see Chap. 5). The assistants' union reverted to NATSOPA. However technological developments were forcing the bookbinders and assistants together again, partly in order to compete with the craft unions for membership in advertising departments. Furthermore, by the early 1980s there were additional problems caused by laws restricting union activity and a worsening economic situation, particularly in relation to import penetration of the UK print market. By 1982 these combined pressures and the failure of the assistants to merge with the craft unions forced the non-craft unions to merge. NATSOPA therefore accepted merger on SOGAT's terms to form SOGAT (82) (Gennard and Bain 1995).

Although fear of the effects of new technology was a crucial factor in forcing craft mergers, in fact initially the:

> Compositor … escaped the impact of technological change … but in the 1970s the impact of technical change adversely affected the compositor, the major occupational group within the NGA. (Gennard 1990: 109)

Subsequently other developments, especially the growth of word processing and the ability to bypass the composing room by sending word-processed material direct to the planning department[17] became the biggest threat. This development meant that many people who were either non-union or not members of a print union could 'produce origination of standards and costs perfectly acceptable to most customers' (ibid.: 118, 121). The union responded by introducing a system of labelling work and refusing to print anything that was not labelled, a system that largely failed, as did the attempt to recruit inplant print workers,[18] a belated attempt to follow the job into a more feminised sector. However, instant printing was treated differently and restrictions were lifted, so these workers could join the union (Craig et al. 1984), a situation analogous to that of machine minders in an earlier period. When taken together with the poor economic situation in the 1970s and early 1980s, increasing import penetration and the increased regulation of trade union rights that limited the ability to take secondary action or continue the closed shop (Fosh et al. 1983; Gennard 1990), the situation seemed dire.

A series of talks led the NGA to merge with the wallpaper workers' union, NUWDAT, in 1979 and later to the final craft merger in 1982 between the Society of Lithographic Artists, Designers, Engravers and Process Workers (SLADE) and the NGA to form NGA 1982 (ibid.).

NUWDAT provided the print union with justification for, and expertise in, recruiting white-collar workers (regarded as a possible pool of alternative cheap labour), as it was an industrial union covering all workers in the wallpaper industry. This was the first real move away from women's exclusion towards segregation, with the establishment of the Art, Technical, Clerical and Administrative (ATCA) trade group whose members paid lower subscriptions. Now there were just three unions in the industry—the NGA (craft), SOGAT (non-craft) and the NUJ (National Union of Journalists).

Meanwhile, for SOGAT the mergers of the 1970s and early 1980s had helped sustain increased membership. Apart from a blip in 1976, membership of SOGAT continued to rise until 1979. In 1982 the amalgamation with NATSOPA increased membership again, but thereafter it fell substantially through the 1980s. The non-manual section established in 1972 was one area where there was success in recruitment up to the mid-1980s and possibly beyond. Certainly by 1985 it was being recognised as the only area where membership was increasing (Gennard and Bain 1995). But the union failed to pursue this route fully.

Such considerations alongside technological, economic and political threats were paramount in forcing a merger between SOGAT and the NGA. Warfare between the two unions provoked merger talks during 1983 and 1984, but from late 1984 to early 1986, talks were effectively suspended. Relations between the two unions were at an all-time low due to the battle over which of them would prevail in the provincial newspaper industry, as technology continued to destroy demarcation lines. The dispute at Wapping with Rupert Murdoch's News International[19] brought them together again with the intention of delivering a speedy amalgamation. This did not happen, in the event, until October 1991 (ibid.).

Technology had a dramatic impact in the newspaper industry, where all the major disputes occurred during this period. But the key effect was that by removing traditional demarcation lines, technology appeared to offer an unprecedented opportunity to SOGAT and the NUJ. They came to believe they could bypass the NGA compositors and thereby gain control of the pre-press areas for themselves (ibid.).

When SOGAT craft jobs were threatened by technology in general print composing rooms, they had suggested sharing the remaining jobs between themselves and the NGA, but the NGA refused. When the NGA was threatened in newspaper composing rooms, they suggested sharing out the jobs between its members and those of the NUJ and SOGAT. Unsurprisingly, SOGAT refused.[20] Both production unions negotiated with employers behind each other's backs, including signing single-union agreements. The NGA continued this practice even after the GPMU amalgamation package had been agreed. Technological change further undermined SOGAT, as unskilled jobs disappeared, either absorbed into craft jobs or passed to new workers, for example, the move to road haulage rather than rail distribution of newspapers and magazines caused a loss of SOGAT jobs (ibid.). Another merger was necessary.

At this point in the narrative, we can see that the primary power resources that male trade unionists relied on were being weakened. Changing technology was undermining technical power, and the state was mobilising legal and positional power to strengthen employers' economic and positional power. Like the rest of the labour movement, the print unions were slow to adapt, thus forcing amalgamation for survival.

Developments Post-Merger: The GPMU

The successful merger between the NGA and SOGAT fulfilled a long-standing policy of both unions to work for one union in print. From October 1991 a new union, the Graphical Paper and Media Union (GPMU), would be the home of organised print workers, an historic development. In the end, pragmatism overcame sectional interest and protectionist hostility. However, by 1993 it was former NGA officials who dominated the union, especially the two most senior positions (ibid.). The result was a tension between increasing centralisation caused by union bargaining weakness and the domination of the union by craftsmen or 'popular bossdom' versus the 'aristocratic' tendency (Turner 1962).

Many pressures confronted the new union, especially increasing merger and acquisition activity and technological change (Gennard and Hayward 2008), leading to a focus on organising from 1994 onwards (GPMU 1997). An analysis of the union's membership revealed continuing falls from 225,000 to 170,000 working members (GPMU 1999). Women made up 18.5 per cent of the membership (GPMU 1991).

However, there was shock when the emptiness of a number of myths that the union believed became crystal clear. Chief among these was the idea that the union had a 'core' membership that would always be there: craftsmen. But in reality, these members were leaving just as fast as semi-skilled members (GPMU 1995: 4). So there was no room for complacency. The information from the Warwick survey[21] showing the average age of GPMU members was 46 and that the proportion of women in the industry who were GPMU members had at least halved in eight years (ibid.: 3 and 6) reinforced this message.

The benefits of the move towards organising were seen in the changing profile of the membership. While a 2004 report on the union's organising experience put the average age of membership at 43 and the percentage of women in traditional areas of membership at 14 per cent, among newly organised members the respective figures were 38 years of age on average and a women's membership of 28 per cent (GPMU 2004a).

Despite some initial organising success, restructuring in the industry with large numbers of company closures overtook this strategy. Membership fell every year of the GPMU's existence, despite organising. Economic pressures piled up: the pound was overvalued; rationalisation through mergers led to super-plants and redundancies. The industry entered recession again in 1999, but this time the redundancies were permanent and came from company closures. And despite settlements in excess of inflation, members began to perceive a loss of union ability to deliver for them (Gennard and Hayward 2008). Consequently, 'branches slowly began to make cuts in their expenditure' (GPMU 2001, 2003, 2004a). By 2001 the union admitted it needed a merger and subsequently succeeded in merging with Amicus in November 2004, despite bitter internal debate over the need to merge and the choice of merger partner (GPMU 2004b; Gennard and Hayward 2008).

The End of Independent Print Unions: Amicus and Unite

This was not the end of union consolidation. In 2007 Amicus merged with the Transport and General Workers' Union (TGWU) to form Unite, currently the largest UK trade union. This is ironic given the bitterness over the previous choice of merger partner—Amicus or the TGWU. The second period of research identified some of the trends in the GPM sector of Unite. Since the global financial crisis, the national agreement had been abandoned. As R2(7) said, 'there has been a lot of overcapacity, there has been a lot of small players going out of the market. There has been a huge amount of administrations and closures'. During this phase there was increasing casualisation, use of migrant workers and chapels selling off various elements of terms and conditions to try and save jobs (R2(8) and R2(6)). R2(6) and R2(7) both identified the warehousing sector as particularly prone to use of agency labour. R2(6) stated that there was really only one large company in print now and one medium-sized one. Technology was still having an impact, despite drastic rationalisation. R2(7) explained that business publishing was now eliding with IT as much of this information was now electronic. However, in conversation with R2(8) during the organisation of another piece of research, she claimed that there were signs of some of the stronger chapels beginning to claw back some of the lost terms and conditions.

Conclusions

The development of the printing industry draws on key primary power resources. Initially coercive state control of all power resources was pursued, including the ability to confer the trade on limited numbers of male workers and even attempts, somewhat unsuccessfully, to control the market for products. For employers economic power will always be of supreme importance. But, of course, they also had positional power, and, in the early historical phase, the state would have conferred the bulk of legal power on them. Public discourse would also have initially favoured employers (symbolic power), while collective power came much later through employer associations.

To counteract these considerable forces, trade unions in print were initially based around craft, attempting to incorporate the restrictive practices the state had found so useful at the industry's inception. Technical power created a commonality among these workers, and the end of the guild, with its challenges to wages and conditions, drew them together as men to form first trade societies and then trade unions (sexual and collective power). These were the foundation stones of the later national organisation of trade unions in printing. The tactics adopted demonstrate an attempt to embed and build on these resources. Tramping and strike pay were both based on strengthening collective organisation and preventing workers with less of a sense of the collective, or who were more desperate for work, undermining attempts to shore up wages and conditions. Some sense of class formation is apparent, with men using rituals and unilateral job regulation to bind them together, alongside an eagerness to adopt the trappings of respectability in the way they pursued their aims.

The gradual success of this approach can be seen in employers' counter measures. To undermine technical power, employers tried to undermine the apprenticeship system, taking on boys before their time was fully served. They utilised 'scabs' to break strikes and adopted piecework in an attempt to increase productivity and control costs. They utilised positional and economic power to further segment the workforce, creating new divisions in the labour process and seeking to embed their 'right to manage'. Finally, they combined to confront the growing success of trade union resistance.

It was the innovation of organising at firm level around the concept of the chapel that, arguably, provided trade unions with their strength at the frontier of control. The myriad rules and practices helped cement a sense of the collective, technical and sexual powers available to members who stuck together. Its relative success would have convinced these men that their sectionalism was their key source of advantage and, evidence suggests, defended it at all costs. This form of organisation had both positive and negative consequences for women as will be seen in the next chapter.

Notes

1. See Glossary for explanation of the three processes.
2. Guilds were voluntary organisations to which members paid contributions and swore an oath. The printing guild was specifically set up to regulate trade and control apprenticeship. Guilds particularly flourished in London, but they had mostly died out by the late eighteenth century (Gardiner and Wenborn 1995).
3. Journeymen were printers who had completed a full seven-year apprenticeship.
4. Trade societies were the forerunners of the trade unions; their purpose was 'to protect their standard of life from the depressing effects of industrialization' (Child 1967: 49) and also provided, for example, sickness and funeral benefit.
5. This was legislation passed in 1799 and 1800 forbidding the formation of combinations of workers aimed at improving wages and conditions. Such combinations were regarded as subversive, especially in wartime (Gardiner and Wenborn 1995).
6. The Call Book was a daily record of unemployed members and 'honourable tramps' who would be sent in order of length of unemployment to fill vacancies notified either by employers or other members. It was kept at the local office or society house (Child 1967: 127–128).
7. Balancing time refers to the practice whereby each working day stands alone and lost time on one day cannot be claimed back from the worker by making him work overtime at standard rates. This remained a rule even in the GPMU rulebook.
8. Progression referred to methods whereby men got access to craft jobs even though they had not undertaken an apprenticeship. They did this by virtue of the length of time they had actually been doing a job already and by serving a period of time in the union after which they were regarded as skilled men. In later periods this continued to be a controversial route to acceptance.
9. The P&KTF was the Printing and Kindred Trades Federation. This organisation was formed in 1901 to provide a means of co-ordinating the activities of the various print unions and to help resolve inter-union disputes. It had no right, however, to interfere in internal union management (Gennard 1990: 12).

10. Employers did not know how to cost and, therefore, charge for work, often making losses, especially where equally ignorant employers drastically undercut each other. Federation would provide experts to help companies produce proper costings for work performed (Howe 1950: 7–8).
11. See Glossary for description of this process.
12. See Glossary for description of these processes.
13. General Printing covers 'packaging, advertising materials, security printing, business forms, books, periodicals, magazines, stationery and catalogues' (Gennard and Bain 1995: 22–23). It consisted of a large number of small firms, some medium-sized firms and a small number of large firms.
14. The British Federation of Master Printers (BFMP), established in 1900, took its membership from companies in general print and later became known as the British Printing Industries Federation (BPIF).
15. The federations were the BFMP and the Newspaper Society (NS) which used to conduct joint negotiations right through until 1980. The dispute in that year led the NS to break away and conduct its own negotiations in order to avoid being caught up in general print disputes. The NS covered employers in provincial newspapers.
16. See Glossary for technical description.
17. See Glossary for description of this process.
18. Inplant print workers were non-printers (sometimes female) in non-print companies operating fairly basic printing equipment. Local government was a common site for this incursion into printing employment. The NGA did try to follow some of this work into the offices but was unsuccessful in convincing the TUC to allow it.
19. The *News International* printed the *Sun*, *News of the World*, *The Times* and *The Sunday Times* originally in Fleet Street but in early 1985 Murdoch moved to a new plant at Wapping, part of London's docklands, without taking his existing workforce with him. The resulting dispute lasted 12 months, and the unions—mainly SOGAT and NGA—failed to gain access to the jobs which were filled by EETPU labour, much of it bussed in from Southampton. The dispute concerned the conditions that would apply to the new technology at Wapping, such as a no strike clause, no closed shop, a different disciplinary procedure for reps and the unfettered right to manage (Gennard and Bain 1995).
20. The NUJ went on to sign a variation on this proposal called the NUJ/NGA Accord.
21. This refers to an internal report compiled by Jeremy Waddington, UMIST, on reasons for union membership and was part of a much larger academic study.

References

Avis, F. C. (1971). *The Early Printers' Chapel in England*. London: Avis.
Bundock, C. J. (1958). *The National Union of Printing, Bookbinding and Paper Workers*. London: Oxford University Press.
Cameron, L. (1967). *Cmnd. 3184 Report of a Court of Inquiry into the Problems Caused By the Introduction of Web-Offset Machines in the Printing Industry, and the Problems Arising from the Introduction of Other Modern Printing Techniques and the Arrangements Which Should Be Adopted Within the Industry for Dealing with Them*. London: HMSO.
Child, J. (1967). *Industrial Relations in the British Printing Industry*. London: George Allen & Unwin.
Clegg, H. A. (1976). *The System of Industrial Relations in Great Britain*. Oxford: Basil Blackwell.
Cockburn, C. (1983). *Brothers: Male Dominance and Technological Change*. London: Pluto Press.
Craig, C., Garnsey, E., & Rubery, J. (1984). *Payment Structures and Smaller Firms: Women's Employment in Segmented Labour Markets*. Research Paper No. 48. London: Department of Employment.
Delafons, A. (1965). *The Structure of the Printing Industry*. London: MacDonald.
Duffy, P. (2000). *The Skilled Compositor, 1850–1914: An Aristocrat Among Working Men*. Aldershot: Ashgate Publishing.
Eisenstein, E. L. (1983). *The Printing Revolution in Early Modern Europe*. Cambridge: Cambridge University Press.
Forster, J. (1956). *Report of a Court of Inquiry into the Causes and Circumstances of Disputes Between the London Master Printers' Association and the London Typographical Society and the Association of Correctors of the Press*. London: HMSO.
Fosh, P., Morris, H., Martin, R., Smith, P., & Undy, R. (1983). Politics, Pragmatism and Ideology: The Wellsprings of Conservative Union Legislation 1979–1992. *Industrial Law Journal, 22*(1), 14–31.
Gardiner, J., & Wenborn, N. (1995). *The History Today Companion to British History*. London: Collins and Brown Limited.
Gennard, J. (1990). *NGA: A History of the National Graphical Association*. London: Unwin Hyman.
Gennard, J., & Bain, P. (1995). *SOGAT: A History of the Society of Graphical and Allied Trades*. London: Routledge.
Gennard, J., & Hayward, G. (2008). *GPMU: A History of the Graphical, Paper and Media Union*. London: Unite the Union.
Gillespie, S. (1953). *A Hundred Years of Progress: The Record of the Scottish Typographical Association 1853 to 1952*. Glasgow: Robert Maclehose & Co.

Gordon, D. (1972). *Theories of Poverty and Underemployment*. Cambridge, MA: D.C. Heath.
Graphical, Paper and Media Union. (1991). Women's Conference Report, p. 1 and Appendices A and 6.
Graphical, Paper and Media Union. (1995). Biennial Delegate Conference Report, pp. 8–9, 91–92, 99–100, 110–112.
Graphical, Paper and Media Union. (1997). Biennial Delegate Conference Report, p. 7.
Graphical, Paper and Media Union. (1999). Finance, Organisation and Membership Committee, Item 5.
Graphical, Paper and Media Union. (2001). Biennial Delegate Conference Report, p. 8.
Graphical, Paper and Media Union. (2003). Biennial Delegate Conference Report, p. 10.
Graphical, Paper and Media Union. (2004a). Report – GPMU Organizing Experience.
Graphical, Paper and Media Union. (2004b). *Direct*, May, p. 2.
Hibbert, C. (1992). *The Story of England*. London: Phaidon Press.
Howe, E. (1950). *The Federation of Master Printers 1900–1950*. London: Federation of Master Printers.
Moran, J. (1964). *NATSOPA Seventy-Five Years: A History of the National Society of Operative Printers and Assistants (1889–1964)*. Oxford: Oxford University Press.
Musson, A. E. (1954). *The Typographical Association*. London: Oxford University Press.
National Economic Development Office (NEDO). (1970). *Printing in a Competitive World*. London: HMSO.
Pelling, H. (1976). *A History of British Trade Unionism*. Middx: Penguin Books.
Phillips, A., & Taylor, B. (1980). Sex and Skill: Notes Towards a Feminist Economics. *Feminist Review, 6*, 79–88.
Porter, R. (2000). *Enlightenment: Britain and the Creation of the Modern World*. London: Penguin.
Sessions, M. (1950). *The Federation of Master Printers: How It Began*. York: William Sessions, Ltd.
Steinberg, S. H. (Revised by Trevitt, J. 1996). *Five Hundred Years of Printing*. London: Oak Knoll Press, Delaware and the British Library.
Turner, H. (1962). *Trade Union Growth, Structure and Policy*. London: Allen and Unwin.
Webb, S. and B. (1913). *Industrial Democracy*. London: Webb, S. and B. for Workers' Educational Association.

4

Missed Opportunities: The Failure of Union Solidarity in the Struggle for Control of the Labour Process

The previous chapter identified how class relations developed within the printing industry and how this was bound up with the formation and reinforcement of male identity. Employers and unions used every tactic at their disposal in these battles. The fluctuating access to power resources acquired by union members affected their ability to resist employers' economic and positional power, demonstrating both continuity and change. This chapter inserts the gender dynamic and will examine how gendered power relations affected women's pay and working conditions.

The early history identifies women's late entry to the industry. Acquiring a better understanding of this period in all its complexity highlights women's relative absence from this powerful and influential industry. From its pre-industrial beginnings, the limitations on women's access to jobs in printing were a key factor. The consequences for women's employment become apparent during the changes produced by industrialisation and the development of the trade union movement through the eighteenth and nineteenth centuries. During this period women's position as subordinate and inferior within the workplace was consolidated and the gendered division of labour entrenched. The twin imperatives of employer strategy to secure cheap labour and trade unions' strategies of resistance

T. Dawson, *Gender, Class and Power*, https://doi.org/10.1057/978-1-137-58594-3_4

reveal a male consensus around the role of women in the printing industry, despite opposing class interests. The rest of the chapter focuses on post-Second World War developments.

The sectionalism already apparent in Chap. 3 is shown to have particularly detrimental effects on women workers. This is examined right through to the demise of the GPMU and beyond to the mergers with Amicus and Unite.

Early History

The key role of the Stationers' Company in embedding work practices was highlighted in Chap. 3. Although the guilds were weaker in Britain than in other parts of Europe, the:

> guild hierarchies and the patriarchal family workshop provided a context in which the father (master of the workshop and head of the household), the mother (masters' wife and helpmate), the son (apprentice guaranteed the mastership after training), the daughter (likely to marry a journeyman), all had their allotted place. (Hufton 1995: 91)

Therefore at the outset relations were already gendered, 'both economic *and* familial' (Lown 1983: 30). Within the guild, men taught men the skill of printing; only occasionally did wives and daughters of master printers become involved in the trade, generally through inheritance (Duffy 2000). In common with many industries, the heavy nature of some of the work served to justify exclusion of women (Game and Pringle 1983; Burchell and Rubery 1994). In addition, 'literacy levels were very low and those of women generally lowest of all' (Hufton 1995), giving men a head start.

The later development of chapels instilled a masculine culture inimical to women. Many customs governed chapel (workplace union branch) behaviour including that the visiting wife of a journeyman had to pay a small sum to the chapel, to which was added more money by the chapel members in order that her 'attendance could be celebrated in the then appropriate manner' (Avis 1971: 35). Such practices, alongside those discussed in Chap. 3 would likely alienate many women. After all, '[i]t would have been an odd family that was willing to see a daughter enter so male-oriented a life' (Cockburn 1983: 19).

In fact women feature little in the early history; the earliest reference, apparently, in 1782 records that a small group of employers tried to use women to break a strike over hours. The union used its recently established campaign fund to pay the women for their support (Child 1967), as also happened with male strikebreakers. However, before 1841 there is little beyond occasional anecdotal examples to indicate that women were employed in the industry (McDonald 1904). Hufton acknowledges that:

> [e]ven when the guilds began to lose much of their power, and governments moved against restrictive practices in favour of a *laissez-faire* economy, women existed as workers most easily in those areas which had never been subjected to guild domination and hence marked out as male preserves. (Hufton 1995: 93)

Women were restricted to domestic power resources, within a patriarchal family, as printing developed. Men had already established technical, sexual and collective power resources, and physical power was useful in justifying women's exclusion from such heavy work. From the outset, their access to power resources was always weaker than that of men, but also operated in tension with class relations, which became apparent in the nineteenth century.

The Nineteenth Century

There was tacit acceptance of women in some bookbinding operations before 1840 and even disputes over their terms of employment. For instance, between 1825 and 1850 a dispute began about women's low pay in bible printing, 6 shillings and 2½ pence for 64 hours (Drake 1920). Braverman (1974) explains how the creation of detail work is an employer tactic to bring greater division of labour. This dispute became especially heated in the 1840s for just this reason (Child 1967). Although the dispute failed to raise wages, it did stop any further pay reductions (McDonald 1904). Nevertheless the bookbinders were the first in the industry to organise women, probably because their attempts at exclusion were less successful than those of the printers. In printing the only early encroachment was of boys and girls who were allowed to undertake the sole task of feeding paper into the machines, a job beneath a skilled man (Musson 1954).

Restrictions on women did not prevent their growing employment during the second half of the nineteenth century, from virtually nothing to 4500 in printing, with a further 14,500 in bookbinding (Child 1967). Very few women encroached on 'men's jobs' and instead filled the jobs created by a further division of the labour process (Duffy 2000). The typographers' and bookbinders' exclusionary practices allowed the formation of a pool of cheap labour, although there was 'no immediate threat to the status of the skilled printer' (ibid.: 131). While the printers continued to organise the pre-press side and latterly the pressroom itself, the bookbinders' unions were initially confined to skilled bookbinders, only later admitting those jobs designated as unskilled and filled mostly by women.

The growth of the Typographical Association (TA) did little to improve women's prospects of getting well-paid work in printing. In the 1880s the union, theoretically, changed its stance and now claimed that there was no objection to women printers as long as they were paid the rate for the job. 'This studied piece of hypocrisy did not deceive anyone' (Child 1967: 189–190) as:

> [t]heir insistence on equal wages was merely a change of front, a cover for their deep-rooted aversion to the entry of women into the trade. They knew well enough that few firms would employ women at men's rates and used the equality argument, in reality, to discourage the employment of female labour. (Musson 1954: 119–120)

The TA was ultimately forced to accept women in printing, if not in membership, as they rose from just 8 per cent to 23 per cent of the sector between 1841 and 1901 (McDonald 1904). Even though some in the union sought to recognise this trend, the union would not organise the women, arguing that if it did it would give de facto recognition to their lower rates, undercutting men's rates too (Musson 1954).

Each technological development caused renewed union opposition to women, for example, the development of letterpress printing from about 1870. Here women and unbound boy labour were restricted to operating the distributing (type) part of the process, but opposition continued even 20 years later when the machinery was almost obsolete (Musson 1954). Still, where they could, employers would use women on keyboards, often at reduced rates, as found in provincial subsidiaries of London companies seeking to avoid the higher rates men earned in the city.

In the late nineteenth century, 'progression' for machine minders became an issue as recounted in Chap. 3. The printers (compositors) tried belatedly to organise them, although this appeared to exclude the few existing women minders (ibid.). These men were usually employed at below TA rates and were generally regarded as printers' labourers (Moran 1964). Acceptance of the apprenticeship system was called into question. The Webbs identified the trade at this time as virtually 'open' because the extent of restrictive practices forced employers to take unindentured labour (1913). Eventually, demarcation between the TA and the assistants' union, the National Association of Operative Printers and Assistants (NSOP&A), was agreed; there is no mention of women but the institution of an apprenticeship system for machine men (Musson 1954). NSOP&A was formed in 1889 as the Printers' Labourers' Union, changing its name in 1899 as a rejection of their designated unskilled status (Moran 1964). This example demonstrates the wholly different treatment of semi- and unskilled men compared to women at this time (Walby 1986). Sectionalism was, thus, maintained by separate union organisation.

There were, however, two experiments in training women compositors that had limited success (Reynolds 1989). The focus on employing middle-class women in distressed circumstances created class hostility with the printers because of the dangers of undercutting, male job losses and deskilling. This only left two extremes for the men—either heavy work that was not skilled or highly skilled work such as imposition (laying out type in pages) that required far fewer people. Such change threatened the violation of the male domain of the case room, a place Reynolds describes as 'uncompromisingly masculine' (ibid.: 37). Although these experiments failed 'employers took the hint' (Drake 1920: 33) and sought to employ female compositors in large numbers in Edinburgh in the late nineteenth century, despite the expectation that they would leave work upon marriage.

At this time, mechanisation of composing equipment required two operators, so employing women was cheaper. Employers were able to exploit women at men's expense. However, employers generally dispensed with women once technology required only one operator (Duffy 2000).

The crunch came in 1872 in Edinburgh. The Edinburgh union lost a strike over the nine-hour day, primarily due to the importation of 'rat' male labour from Liverpool and London (Reynolds 1989). In the middle of the three-month dispute, the employers brought in literate young women, who were attracted by the comparatively high wages, to be trained as compositors. The women were accused of strikebreaking (McDonald 1904; Webbs 1913; SOGAT 1985: 6), playing down the key role of skilled, male strikebreakers and perpetuating the myth of women's lack of solidarism (Barron and Norris 1991, originally 1976). Although it does not excuse them, given the women's lack of training, their use as strikebreakers would have been severely limited compared to the skills of male strikebreakers (Reynolds 1989; Duffy 2000), but the greatest vitriol was reserved for women.

The strike seriously weakened the union and allowed women compositors' numbers to increase until 1910 when the union felt able to challenge their employment, especially on the new composing machines. The employers resisted this challenge, and the women were split between class loyalty and their need for a decent wage. About half the women (400) opposed the men, partly incensed by the men's suggestion that they join the unskilled union in England. Men claimed that middle-class suffragists or the employers were provoking the women's opposition, and rumours circulated that women intended to carry on working if the men struck (Reynolds 1989).

Some commentators ignore the women's resistance to being forced out of work (Zeitlin 1985; Duffy 2000), perhaps because of its failure. Ultimately the strength of the men's organisation prevailed, and women were forced out of the trade over a period of years (Reynolds 1989). Similarly, in England, the bookbinder's union aimed to limit women's employment until they could eliminate them from craftwork altogether (Bundock 1958).

Further disputes occurred in the 1890s in Glasgow, Dublin and Manchester about women's employment (Bundock 1958). Although it is not explicitly stated, it appears that in both the Glasgow and Manchester cases, women were not forced out but were restricted as to the type of work they could do. It should also be noted that these disputes occurred in the wake of the Taff Vale judgement,[1] but this did not deter the men from taking action, although the national union tried to avoid these disputes. In a further example from 1893, the London Journeymen Bookbinders even drew up an elaborate agreement, without consulting the women, about what jobs women were entitled to do (McDonald 1904).

Union views were split on how to handle the encroachment of women into the industry. The Bookbinder's chief secretary said:

> How is it, when some kinds of labour pass from men to women, that the value falls nearly 50 per cent? Is it a just thing that women should be paid 10/- or 12/- for doing work that is worth 20/- when done by men? That is simply a matter of might and right; and its obvious explanation is the fact that women are not organised, and therefore unable to defend or uphold their rights. (Bundock 1958: 67)

But not enough men were ready to hear this message. A compromise solution was to organise women in their own unions in order to reduce employers' ability to drive down wages (ibid.), preserving gendered sectionalism. Meanwhile lithographers (printers) actually preferred girls to boys for feeding the machines because 'no girl attempts to rise into the higher grades and "pick up" the trade without apprenticeship' (McDonald 1904: 11).

While there was one unsuccessful attempt at women's organisation in the Society of Women Employed in Bookbinding (Bundock 1958), greater success was achieved with the Manchester and Salford Society of Women in the Bookbinding and Printing Trades, formed in 1896 (later merged with the NUPB&PW in 1942). In its 46-year life span, the union improved women's terms and conditions, expanded its geographical coverage and employed a full-time woman officer (ibid.). While often working in tandem with the men's unions, there was also friction over male bookbinders' petty restrictions on the work women were allowed to do, for example, 'the pasting of end papers—a process "within the capacity of a child"—must not be done by women at all' (Drake 1920: 153). However, the women agreed with the men that they should get the rate for any job they undertook but refuted implications that they were not capable of doing skilled work and made clear their willingness to be trained.

Meanwhile employers were split. On the one hand, they did not want strong union organisation among women, but, on the other, they wanted to prohibit undercutting between companies that arose from the ease of cutting unorganised women's wages. McDonald (1904) reports that on the women's part they mistrusted unions and thought they would be ineffective in bargaining for women, that membership was risky and the proximity of marriage made membership pointless as they would likely give up work. In addition, job segregation prevented contact between men and women, so the men often did not think it worth organising women (ibid.).

In 1889 the Printers' and Stationers' Warehousemen, Cutters and Assistants' Union was formed (Bundock 1958). It signalled the arrival of 'new unionism' in the industry to take account of the craft unions' failure to organise semi-skilled and unskilled workers. This union had less of a problem with its attitude to women and helped organise the Women Bookfolders' Union in 1894, apparently the forerunner of the London Women's Branch that appeared in later years. A boost for women's organisation was at last in sight.

Developments in the nineteenth century demonstrate how women were 'on the back foot' from the moment they entered the industry, sectionalised by skill. Men's already established and growing technical prowess provided scope for their increasing sexual and collective power as they organised in 'men only' unions. Women were denied all these resources initially, only grudgingly accepted for the most menial tasks and blocked from any alternative avenue. Organising them into women's unions mainly compounded their weakness because of their limited access to technical power, any increase in visibility wasted. The employers were only too happy to exploit them and wait for technology to help them use this pool of cheap labour to undermine men's organisation. The question remains: why did men take this route? Although employers were looking for opportunities to use women for the benefit of capitalists, leaving the women vulnerable made this easier. The class struggle was thus undermined by exclusion. If women had entered the industry, been trained and organised, why would they not have wanted and fought for exactly what the men had—better wages and conditions? This account suggests that gender relations interacted with class to ultimately weaken both men and women workers, through embedding gender sectionalism. The twentieth century saw this journey brought to its logical conclusion.

The Early Twentieth Century

The growth of print unionism led to more disputes. Some of that unrest emerged from women's activism. For instance, in 1902 women and men struck against Wyman's Reading and London plants, resisting attempts to introduce both male and female cheap labour (ibid.). Also contrary to male printers' beliefs (Duffy 2000), McDonald reported that married women tended to agitate to protect pay rates as they 'generally have more than themselves to keep' (McDonald 1904: 106), while unmarried women living with their parents do not have the same living expenses and may accept reductions. In fact many employers saw married women as 'the centre from which general discontent in the workrooms emanates' (McDonald 1904: 107).

Training for women was a controversial issue at this time given that employers had no set system and often used women to undermine the piecework system to cut costs. Technology reduced opportunities for women learners (a form of women's apprenticeship), and where large machines replaced women's work, employers thought women were incapable of running them, except sewing machines. Therefore '[n]o process ... seems to have been opened up to women in consequence of the introduction of machinery' (ibid.: 96).

Meanwhile the non-craft Printers' Warehousemen and Cutters merged with the London-based Women Bookfolders' Union in 1903 that they had helped create in 1892 (Bundock 1958). Agreeing to admit only women who were not engaged on work traditionally done by men allayed the male membership's fears. From 1910 women appeared on the non-craft union's executive council and amongst its organisers. The union had appointed a woman organiser in 1906 and by 1908 had 1000 women members out of a total of just over 4000 members. At this point the women's branch got its own administration, headed by a man.

By 1911 all bookbinders and related workers were in one union, but the Manchester and Salford women's union was left out (ibid.) even though women's employment growth rate in the late nineteenth century had been more than double men's (Child 1967). Opponents of women's organisation declared that organising women would legitimate their work and that '[f]emale labour is competitive and where female organisation exists it will be competitive. The women will fight male labour wherever it is possible' (Bundock 1958: 86), a comment somewhat at odds with the evidence.

The union was unable to remove women from the industry but could not conceive of organising them, so further disputes occurred, such as that in Scotland in 1913. The members felt that while there were unemployed men on their books, the employers should not take on women. Conversely, the employers' alliance argued that skills' shortages necessitated an alternative labour source and the simplicity of the work would suit women. Six hundred members of the bookbinders' union went on strike for over a month. The resulting agreement provided that where machines were being brought in which would displace male labour, the operative's wages would be protected and only those with a seven-year apprenticeship would be given 'first claim' to work them (ibid.: 94). So there was no new strategy. Wherever men identified themselves with a craft, women were subject to segregation at best, exclusion at worst.

Two World Wars and the Interwar Years

The advent of war was met with attempts to ensure women's exclusion, but eventually the TA agreed a limited relaxation of the rules on condition that it 'be regarded purely as a war emergency, and pre-war conditions shall be reverted to when the war is over' (Musson 1954: 367; Howe 1950). Employers were to utilise apprentices and retired members and relax demarcation rather than allow women dilutees[2] on machines, on pain of industrial action (Child 1967). This occurred despite continuing labour shortages that raised the men's wages. Generally, the union was successful in holding the line. Once again women lost out in the wrangling over their access to jobs, the men still seeking their exclusion and trying to exploit labour shortages to their own advantage, while the employers were accused of wanting cheap labour (Musson 1954). However, the use of women as substitutes was limited by the need for proper training and the requirement to pay the rate for the job (Howe 1950). In bookbinding, women's encroachment was treated with equanimity as there was not 'the slightest fear but that when the men come back the women will be only too pleased to vacate the jobs they have taken on whilst the men were away' (Bundock 1958: 190).

Records indicate that even during wartime, when there was runaway inflation, women received much lower wage rises than the frequent and substantial rises men were awarded (Bundock 1958). Given that these wage increases were less than inflation, this would have had a disproportionate impact on women. Conversely, in the 1920s, the unions were forced to accept wage cuts and the men's were greater than the women's (Musson 1954).

Women organisers achieved considerable influence in the non-craft union because many of the male organisers were called up or volunteered (Bundock 1958). In 1913 the non-craft union became the National Union of Printing and Paper Workers (commonly, the Paper Workers) and continued to organise women whose potential membership was substantial. For instance, in Aberdeen in 1917, union assistance in a women's dispute led to the recruitment of over 3000 members.

Meanwhile the bookbinders had barred women from their trade union and, rather disingenuously, changed their stance when they realised that there were between 14,000 and 15,000 women occupied in jobs that did not encroach on 'men's work'. Subsequently, they quickly organised in excess of 9000 women at a time when they only had 8500 men in membership. This violated the Paper Workers' traditional territory and led to a crisis in relations between the two unions that came to a head in 1917. The Paper Workers Union thought the bookbinders were simply trying to steal its members and potential members.

The bookbinders' union appears hypocritical in claiming that they had only been waiting to complete the existing merger potential before addressing the question of women. This contention was at odds with its attitude to women, as demonstrated in one of its constituent unions in 1894. A member was summonsed to explain why he had accepted a position in a company where there was a woman employed as a ruler. He said he had taken the post with the express purpose of removing the woman and he had succeeded by setting her a task, without any guidance, and sacking her when she failed, as he knew she would. He was commended 'for his diplomacy and discretion in the matter' (ibid.: 297).

Similarly there were questions about just how successful the Paper Workers Union actually was in recruiting women. Stalemate had been reached. The bookbinders' union was expelled from the P&KTF, and the Paper Workers threatened to add 'bookbinders' to its constitution in order to allow it to retaliate and start organising in bookbinding departments (ibid.).

The deadlock was broken when the Paper Workers issued an invitation in 1919 to have merger talks with the bookbinders. The realisation that they could not continue at loggerheads led to amalgamation in January 1921. The National Union of Printing, Bookbinding, Machine Ruling and Paper Workers (NUPB&PW) was born over the conflict generated by the failure to organise women. It had over 100,000 members of whom 60,000 were women (Child 1967). The executive council of the new union contained seven women out of its 26 members. This was undoubtedly a step forward but did not represent the proportion of women in membership and could be construed as incorporation, outflanking potential demands for greater representation (Cunnison and Stageman 1995; Clegg 1989).

The development of centralised pay negotiations did not bring much relief for women. Their pay remained separate and was determined according to 'age and length of experience in the trade' (Howe 1950: 71). Some indication of the size of women's employment at this time can be estimated from the new Joint Industrial Council's figures revealing employment of 170,000 people, of whom 79,000 were women. Although simplified printing equipment in offices and the use of women as readers threatened greater encroachment on male employment, little happened to increase women's union organisation (Musson 1954).

In 1918 the TA had reaffirmed its policy of excluding women from craftwork, often securing agreements to remove them through natural wastage and prevailing upon non-craft unions only to organise women for unskilled jobs. This was not unusual (Cunnison and Stageman 1995) but precluded any large influx of women. This can be contrasted with the more relaxed attitude to boy labour that did not serve an apprenticeship but by 'progression' and five years' union membership, working at the trade, qualified for membership of the TA (Musson 1954).

The printers' opposition to women was almost total, going so far as to scupper a proposed amalgamation with the Scottish TA because it recognised women compositors. Even the benefit system was utilised in the pursuance of this policy. During the period that the TA was responsible for administering state unemployment benefit, it often appealed against denial of benefit for men who had refused to work with women (ibid.). Meanwhile, for the employers cheap labour was not always a blessing. The problems of organising the industry especially in relation to small firms meant the employers' federation 'complained bitterly of cutthroat competition from houses … [*inter alia*] employing cheap female … labour' (ibid.: 407).

Union hostility did not prevent women striking in solidarity to oppose pay cuts alongside the men in the NUPB&PW (Bundock 1958), but the men still continued to challenge every instance of women doing 'men's work'. Just such an instance occurred in 1922 in Manchester. The men struck over women's encroachment, and the women, who were in the Manchester and Salford Society of Women in Bookbinding, did not support them. However, it is recorded that eventually a settlement was reached that was satisfactory to the women but did, indeed, remove them from the offending jobs. Even when men did, ostensibly, support challenges to women's exploitation, it was often just to use them as a bulwark. For instance, an attempt to reduce women's rates in London manufactured stationery work was only resisted because the union believed the subsequent arbitration award had deterred employers from seeking a general reduction in rates (ibid.).

Not even small concessions were supported, such as the 1930 proposal for reserved seats for women on the NUPB&PW's executive council, which was defeated. Even in its own London Women's Branch, it was not until the male official died in 1937 that a woman secretary was appointed (ibid.).

The Second World War called for the relaxation of the rules again. While the bookbinders were less stringent, the printers allowed in women dilutees only under strict conditions. The printers' executive council had to sanction each case and only allowed women in when every other source of labour had been exhausted (Typographical Circular, May 1941a: 82–83). In addition existing members were assured that, unlike in 1914–1918, 'much firmer guarantees have been given of their removal after this war' (Typographical Circular, September 1941b: 137). Women were to be kept to the less demanding work to ensure an easy transition back to the pre-war rules when hostilities ended but paid the full rate (Musson 1954; Child 1967). However, there were still labour shortages given that at the outbreak of war the industry employed about 270,000 people but by 1943 108,000 had enlisted (Howe 1950).

The first half of the twentieth century saw entrenched hostility to women and every possible tactic used to prevent them accessing better jobs or increasing their pay relative to men. This period saw trade unionism consolidating its organisation, women were eventually brought into some of the men's unions because of their sheer numbers, and a limited form of sexual and collective power was realised and limited level of visibility. But their representation in the unions did not reflect their proportion of the membership, and the few significant positions on offer could be seen as a form of incorporation to ensure their sexual and collective powers remained muted. Maybe their earlier suspicion of the unions was justified. But examples of women's militancy raise further questions about their oppression. If they were so willing to support striking men and there was evidence of married women, in particular, being more militant on their own behalf, why was it necessary to restrict them? Clearly women were prepared to fight, so why not build solidarity through demanding proper skills training and access to jobs for all workers, preventing the use of women as cheap labour? Again it seems gender relations trumped the building of cross-gender working class solidarity to resist the employers.

Post-Second World War

In the 1950s general printing 'employed over 22,000 women, most of them ... [bookbinders' union] members with few skills on hand finishing work' (Gennard and Bain 1995: 10). Although the bookbinders' membership was 44 per cent women, by 1955 there were no women national officers or regional secretaries, minimal representation on the executive and among the organisers. The London Women's Branch contained all the women members in the city, and throughout the country single-gender chapels were common. As with other sectors (TUC 1955), women were absent or segregated, effectively invisible as they had been since before the war.

Learnerships for women were still common, as R2(2) explained, '[you did] three years [training] and two years' experience money before you got full rate'. She also described the post-war print industry. When she got married in the 1950s, she had to leave and go jobbing (casual) because some firms operated a marriage bar. She was jobbing for the next 20 years. And, although she did not concur, she said many women accepted that men were the breadwinners and you should not interfere with their jobs. She was in the London Women's Branch and explained how their branch secretary, Rene Chubb, was a real fighter who pushed women to stick up for themselves. 'They didn't always want it and they didn't always thank her for it but she was one hell of a fighter', a view which R2(1) and R2(9) strongly supported. This included pushing for access to overtime and fighting the Factory Act restrictions on women's hours, as well as campaigning for the Equal Pay Act. Furthermore, when R2(2) got onto the EC, other women, outside of their rather insular London branch, became visible, and it opened her eyes to their shared problems. But as far as skilled men were concerned, her experience was that they 'resisted women every way they could'. Conversely, the union organisers, especially the late Maurice Suckling, were great allies generally and helped win the greatest gains for women.

The formation of SOGAT appeared beneficial to women members as, although only two women out of 24 delegates sat on the National Executive Committee (NEC), the 1968 Delegate Council established a

Women's Advisory Committee (Gennard and Bain 1995). Opposition was based on women's failure to stand for NEC elections and the dangers of 'creat[ing] a union within a union' (ibid.: 264). In the event, there was only 30 per cent turnout in the committee ballot, and it met a total of only five times during its life. Unsurprisingly, the committee folded in 1972 partly, Gennard and Bain claim, because its members 'found it difficult to segregate men's and women's interests'. The general secretary did not believe that 'there are many exclusively women's problems … [anymore] than there are exclusively men's problems' (ibid.: 268).

Despite these setbacks women continued to fight for their place in the industry and union as they did in 1976 at Format International, a London security printer with 10 male employees and 20 female employees in separate SOGAT branches. The dispute was, ostensibly, over demarcation at a time when work was fluctuating: what was the house custom for the weight that women should pack[3]? The women were members of the London Women's Branch (LWB), and the men were members of the London Central Branch (LCB). The men walked out and were dismissed, and only a few ever reinstated at the end of the two-year dispute that saw almost a complete breakdown of relationships between the LCB and the national union. However, both the NEC and the Final Appeal Court of the union heard evidence that the men were trying to take work from the women so that they could continue to get overtime. The women were left with little or no work during the trial period they acceded to this request and they decided to take the work back. In the words of one witness to the appeal, 'since the men walked out the women had been subjected to vile verbal abuse such as … "why don't you lie on your backs, earn more money?"' (SOGAT 1977: MSS.39/SO/3/9/3)—one of the mildest rebukes recorded. As the company was local, the abuse went on outside of work.

When the women were asked why they did not obey a union instruction to support the men's unofficial strike, they said it was because they had no alternative work, while the strikers were claiming social security benefit as well as working as casuals on newspapers. One witness to the NEC meeting where the women were to be disciplined explained that they had asked whether the union could get them work if they walked out in support of the men and were told they could not guarantee it.

The NEC report recorded, 'She could not express the feelings the women had in knowing the men were still employed and yet, if the women stopped work they would have no jobs at all ... As Mrs Chubb had said, the women felt that since they had been in the union they had had no good come of it at all' (ibid.). The men justified their walk out on the grounds that they were breadwinners. But as the women said, many of them were breadwinners too. Sectionalism was pursued using physical, sexual and collective power resources in what appears to be a 'land grab' at the women's expense. Evidence from R2(1), R2(2) and R2(9) suggested such 'land grabs' were fairly common, and it was only assistance from supportive organisers that alerted them to resist such moves. However, R2(9) captures the class/gender tensions well in her disappointment that the Format women would cross a picket line.

Meanwhile the introduction of typewriter keyboards to the industry caused real concern to the NGA, especially when employers started to employ women operators on 'the minimum BPIF keyboard rate rather than the keyboard rate for the area' (Gennard 1990: 110).[4] This meant that women keyboard operators were cheaper than men. So, the Equal Pay Act 1970 was seen as helpful in ensuring occupational segregation because it made women more expensive, thereby assisting in maintaining 'an all-male presence, the implications being that all ... things being equal the employers will opt for a man rather than a woman' (NGA 1976: 24). This is an example of how well intentioned public policy can be short-lived (Phelps Brown 1977), and here the result was to reinforce women's invisibility.

A 'belt and braces' approach saw the NGA insist that employers advertise in the national, local and trade press and the union journal, employ male telex operators or disabled men (NGA 1976: 24). Only when these routes had been exhausted would the union countenance giving women a dispensation to operate keyboards (Gennard 1990). Union members' reluctance to operate keyboards and labour shortages in the industry (National Institute of Industrial Psychology 1969) undermined this approach. The culture of masculinity was also implicated in that the NGA condoned for many years the objectification of women with 'platemate of the month' (photographs of women in bikinis) in its journal in the 1970s, and trade advertisements portraying scantily-clad women draped over printing machines until the early 1980s. The 'banging-out' ceremonies for apprentices (see Chap. 3) also continued until quite recently (NGA 1988: 6–7). So twin strategies of exclusion and segregation continued to be pursued in a culture of masculinity, sustaining women's invisibility.

Women were around one-third of SOGAT's membership (Gennard and Bain 1995) and the election of Brenda Dean as general president in 1983 and then general secretary in 1984 (Dean 2007) focused attention on their representation. This was the first time a woman had held either position. Subsequently, women members' profile improved with the adoption of a women's page in the union journal and the setting up of a Women's Positive Action sub-committee in 1983 (SOGAT 1984). The union also commissioned a report *Women in SOGAT (82)*,[5] which was debated at the union's 1986 conference (Gennard and Bain 1995). The impetus had come from the TUC Charter *Equality for Women Within Trade Unions*.

Women in SOGAT (82) exposed the lack of representation for women in the union. Only one of ten national officers was a woman, none of the 18 national organisers or group area secretaries were women, and there were only two women on an NEC of 36. There were four women out of 93 branch officials, and 38 out of 417 delegates to the union conference were women. At chapel level women were 26 per cent of chapel officials where there were more than 50 members and 21 per cent in chapels with less than 50 members.

Local initiatives were patchy, and although there were many women, activists were 'not in the most influential union leadership positions' (Ledwith et al. 1985: 49), even at branch level. The report did not recommend that the union appoint/elect a women's officer but that women's equality should be mainstreamed into existing committee structures through the:

> establishment of Equality Working Parties at every level in the union and the appointment of Education and Training Officers and Equality Officers at every level as well as reserved seats on all committees. (Gennard and Bain 1995: 488)

However, the NEC rejected the proposals, which were merely noted by conference with no real debate, relegating it to the role of a discussion document for branches (SOGAT 1986). Men's collective, positional and personal power resources were mobilised to marginalise women, focusing instead on the Wapping dispute.

Even the request to consider the establishment of a national equality committee was opposed by the general secretary, Brenda Dean, because it would need a different structure to other committees (ibid.). The only successful motion supported extra women's training courses (ibid.). Even at this late stage, structural change was being resisted, maintaining women's invisibility.

Brenda Dean is a complex figure in this situation. While most women continued to be largely invisible, she was anything but. Brenda Dean's account of her life in the union reveals the complexity of her position and how she attained such high office. In essence, it appears that in both the ballot for president and general secretary, the large number of male opponents merely succeeded in splitting the men's vote (Dean 2007). R2(1), R2(2) and R2(9) believed the splitting of the London vote was crucial, allowing Brenda's northern constituency to get her elected.

Interestingly, Dean also claims that she 'was a trade unionist who just happened to be a woman and … was interested in their [women's] issues … only as part of … [her] interest in all working issues' (ibid.: 45). However, later she concedes that 'I had to contend the top job, not because I particularly sought it but because I owed it to all the women in the union who had so clearly backed me for president' (ibid.: 52). It is also clear that she recognised women's low pay and status. Nevertheless she was divisive even among women activists and officers. While two women officials (officer 4 and R2(7)) explicitly found her supportive and encouraging, one activist (R2(2)) thought she was a good GS generally, and R2(1) and R2(3) harboured considerable anger about her role in the Wapping dispute. R2(3) commenting on the first GPMU women's conference workshop run by Brenda found her approach to equality disconcerting, 'I can't remember what the workshop was supposed to do but what it appeared to be from her point of view was tips on how we should dress and present ourselves, which was a bit alarming.' R2(4) and R2(2) gave a more nuanced view of her position. The former indicated that there were times when she stood up for women, for instance, on access to cancer screening at work, but 'not always when a tougher stance was needed … She always fought for equality but in terms of being identified in line with a cause, perhaps not'. While R2(2) pointed out that she was always heavily outnumbered by men on the EC and so would have had difficulty pushing through anything too radical, a typical constraint on women leaders (Connell 1983).

One issue that was tackled was the continued existence of single-gender chapels. These had been criticised in a tribunal decision involving the company Thomas De La Rue of Gateshead. They were seen as having perpetuated unequal pay and lack of job mobility for women, thereby reinforcing job segregation (Gennard and Bain 1995).

But the discrimination laws that were supposed to assist women in achieving equality actually backfired on the London Women's Branch (LWB) in 1987. The branch had three full-time officials and a committee of twelve and was the second largest branch in the union (Nicholson 1994). They were vigilant in protecting women's interests, but sometimes women's reluctance to accept positive action could hinder progress (ibid.). This single-gender branch problem had surfaced in the 1970s, even before the Format International dispute, and became somewhat entangled in discussions in the dispute. Subsequently, the Equal Opportunities Commission (EOC) had concluded that the system perpetuated job segregation so it issued a non-discrimination notice. The branches were forced to merge, although R2(9) claims that the EOC chair denied this and it was actually the union that forced the merger to rationalise the costs of two branches with falling memberships. But the men outflanked the women, deciding committee places before the merger became formal, leaving women in a weaker position (R2(2)). Worse was to come. Of their three officers, two retired soon after the GPMU merger, leaving one (R2(9)). She came up for re-election and was opposed by the male EC representative who sat on the national women's committee. He won because of the internal politics of post-merger positions and bad feeling in the branch (R2(2)), leaving not a single woman officer in the merged branch (GPMU 1994b). Where women had progressed they now sank back into invisibility again, men's positional, personal, sexual and collective power overwhelming even the most strongly organised women.

Demographic trends and technological change meant that there were more women in the sector; the drawback was that most were in white-collar occupations (Gennard and Bain 1995). SOGAT changed some of the contribution conditions and employed two organisers. White-collar workers were in the CMS branch, which had a considerable proportion of newspaper clerical and managerial staff in membership (interviewees R2(4) and R2(7)). But the actions taken were, in the end, inadequate having little impact on women's recruitment (ibid.).

The NGA tackled white-collar workers by setting up a trade group for them. Other changes included the scrapping of the examination that had to be sat before a member could become a candidate for national office. But this did nothing to change the profile of national officers who remained exclusively white and male. The only woman ever to reach national office in the NGA gained her place as a result of the amalgamation with NUWDAT and not by election. However, the NGA was always hostile to any attempt to introduce appointment of officers. Gennard refers to two occasions where the issue was raised at delegate meetings but was defeated in favour of retaining elections.

Clerical jobs had increased by 9.5 per cent in the 1980s, whereas there was only a marginal increase in employment for women working in traditional printing jobs (ibid.: 257–258). So the print unions' continued failure to follow the job where the areas were more likely to be feminised, for instance, inplant printing and clerical work, left them vulnerable. The desire to preserve sectionalism meant women remained largely invisible and irrelevant to workplace bargaining.

In 1990 there was an important development when SOGAT agreed to hold an annual one-day conference for women members, the result of a protest by women delegates at the union's biennial conference, to discuss equal pay, opportunities, women's rights and positive action in the union (ibid.). The first conference took place in Manchester in 1990 and was attended by over 100 delegates. This was an opportunity for women to raise their visibility.

Demographic trends, new technology and the need to expand into new areas meant even the NGA must 'change its image of being an anti-female union' (Gennard 1990: 452). In particular, any future amalgamation would almost certainly take them into a union with far more women members. However, the union remained a white, male monoculture. Its female membership had grown from 203 in 1972 to 8000 by 1991 (Gennard and Bain 1995), but this was still a small minority in a union claiming 130,000 members. For the NGA the 'culture shock' of amalgamating with a bigger union that was not predominantly craft-based would be considerable.

The NGA's experience of representing women members entering the industry through the utilisation of new technology was not encouraging. Throughout the 1970s and 1980s, there were a number of disputes in the provincial newspaper sector. They often involved the bypassing of the composing room (pre-press) through the use of computers operated by women (Gennard 1990). By the late 1980s employers were keen to derecognise the NGA (Dawson 1993). They undermined the principle of equal pay by levelling down instead of up, reducing men's pay not increasing women's. For instance, the *Oxford Mail* encouraged highly paid men to leave on enhanced redundancy terms or face reduced rates while recruiting an alternative (often female) workforce on 'rates some £100 a week lower (than existing rates]' (Gennard 1990: 500). Unsurprisingly, many compositors took the money and ran, so rates automatically fell drastically in the sector and de-unionisation became widespread.

Employers were also concerned about demographics. This led to a joint NGA/SOGAT/BPIF initiative, called 'Achieving a Balance'. It was aimed at encouraging employers to proactively adopt procedures for recruitment, promotion, selection and training that would lead to a more gender and ethnically balanced workforce (BPIF 1988). Such 'proceduralisation' had little effect, and attitudes remained largely unchanged (see Chap. 6).

From the 1980s onwards, there was greater inclusion of equality issues on the bargaining agenda, such as childcare (GPMU 2003), maternity/paternity benefits and sexual harassment, but mostly with little success (Gennard 1990). Neither was there any progress on training, despite the demise of the apprenticeship system, the effects of demographic trends and the NGA policy to achieve a 50/50 intake of male and female trainees.

Even the NGA women's committee suggested that women were at fault in not applying for skilled jobs because they did not want to get their hands dirty (Gennard 1990: 448), perhaps indicating the degree of incorporation of attitudes of the few women who had entered the hierarchy. Employer resistance and members' ignorance of its provisions undermined the national agreement's effectiveness (ibid.). SOGAT suggested to its NEC that even industrial action might not be sufficient to achieve these policies (Gennard and Bain 1995).

Reflecting on the post-war period, R2(4) felt that much of what SOGAT men did was kindly meant but paternalistic. While on the other hand women were habitualised to their positions in the industry—as she had claimed in an earlier conversation with the author 'they're paid Class 3, treated Class 3, act Class 3'.

Technological developments, dissolving demarcation lines and a host of other pressures also forced the print unions to engage with the wider labour movement and provided scope for increasing women's visibility. No longer isolated, the print unions' anachronistic behaviour towards women and minorities came increasingly under scrutiny through greater involvement with the TUC, which provided an avenue for women's participation (ibid.: 374, 376). Both the NGA and SOGAT were committed to the TUC Charter *Equality for Women Within Trade Unions*. However, it only resulted in limited action given that, for instance, the NGA could find 'no formal barriers [to women's participation]' (Gennard 1990: 446). It took the merger of the NGA and SOGAT to move this position forward significantly.

During the post-war period, the unions were forced, however, reluctantly to make room for women, increasing their visibility. But right through the significant changes in technology during the 1970s and 1980s, they continued to practice exclusion and segregation to the greatest extent possible. Only the growing confidence and experience of women trade unionists, the resulting changes in the law and peer pressure from the wider labour movement began to permit women to slowly gain some positional and personal power to enhance their limited technical, sexual and collective power resources.

The GPMU Merger

Women's participation in the union's structures improved in the GPMU as, strategically, the union had to deal with the potential centrality of women in both a changing labour market and process. Women constituted about 18.5 per cent of the union's membership, having been 8 per cent of the NGA's membership but 27 per cent of SOGAT's (GPMU 1991). 1992 was a watershed year for women in the union. The women's committee submitted proposals for women's representation on the EC (GPMU 1992b), which were accepted. The paper made a strong case for women's representation on the EC, citing demographic trends; the appalling level of women's representation at all levels of the union; the requirements of other bodies, such as the TUC and Labour Party, for proportional representation; and, crucially, the failure to recruit and organise women (GPMU 1992b). The TUC had re-launched the Charter *Equality for Women Within Trade Unions*, specifically calling for proportionality (Cunnison and Stageman 1995). Union density itself was disproportionate, with 56 per cent among male workers but 'less than 22% amongst female workers' (GPMU 1992b: 9). At no level did the union's decision-making structures contain more than a handful of women. Furthermore, the only remaining female national officer and a number of male officials also left, leaving the union dominated by male, former NGA officials.

The subsequent proposals called for guaranteed proportional representation (GPR). This approach incorporated a number of key features. In particular, the union wished to avoid tokenism, to use its existing electoral system, proportionality, in determining the necessary numbers to be elected and the greatest possible legitimacy for the successful candidates. But even then there was an attempt at the union conference to draw the women for the EC from those standing for the women's committee. Many women spoke forcefully against this proposal, and it was defeated (GPMU 1995). The 1995 union conference introduced the GPR rules covering the executive council, chapel and branch committees and national delegations (ibid.). To meet these aims, the procedure finally adopted for the EC required women to stand in the normal election process, but if insufficient women were elected in their own right to meet the requirements of proportionality, then the women with the highest proportion of votes in their region would be deemed to have been elected (GPMU 2000a). In the 1996 election (the first using the new formula), 11 women were nominated, the highest ever number, and 4 were elected in their own right (GPMU 1996a).

The merger talks established the women's committee in the rulebook, consisting of ten women elected on a regional basis and three executive council representatives (GPMU 1990). Later a biennial women's conference also entered the constitution (GPMU 1995). But it still took until 1996 to ensure the women's committee was fully female (GPMU 1996b) and until 1999 to replace the (male) president of the union with the women's committee chair as conference chair (GPMU 1999a). However, the women's committee was prevented from discussing aspects of the national agreement for some years (Colgan and Ledwith 2002: 172). Recommendations from the conference went, via the women's committee, to the EC for endorsement. Contrary to more advanced forms of self-organisation[6] (Munro 1999), both the committee and conference were linked to the decision-making structures of the union.

However, one woman, in particular, received a setback. Tony Dubbins, the former NGA general secretary, defeated the high-profile SOGAT general secretary, Brenda Dean, in the ballot for general secretary of the new union, despite her being the leader of the larger of the two unions. She subsequently left the union. Gennard and Bain claim a number of factors affected the ballot's outcome, causing hostility to Dean, including the 'concentration … [in recent years] … on increasing the pay of lower-paid members to the detriment of craft differentials' (Gennard and Bain 1995: 244), the lack of 'industrial response' to the NGAs encroachment on their (SOGAT) jobs and the 'domination' of the union journal by the 'Women in SOGAT' section (Gennard and Bain 1995). However, the latter was contested by R2(9) who felt Dean's politics and the Wapping dispute were among the most salient reasons and Dean admits to her 'New Labour' tendencies (Dean 2007). Furthermore, Gennard's comment on differentials is at odds with R2(4) who said, 'it was inherent in SOGAT politics to fight for lower paid workers'.

GPR made less of a difference than expected. The first executive council elections were a great success, but subsequent elections saw women's representation fall back again. Only four women (one using the reserved seat mechanism) out of nine nominations were elected in 1998 (GPMU 1998a) and only three in 2001, the total nominated (GPMU 2001a), leaving GPR requirements unfulfilled. Explanations have not been forthcoming for this lack of candidates. GPR was seen as having been delivered by the general secretary and not by women's campaigning, even though they had to win his support (Colgan and Ledwith 2002: 171). There was also something of a backlash after the first successful elections, with threats to take away GPR, as it clearly was not needed (ibid.).

Experience of GPR at local level, where it could most touch the lives of women members, was only ever patchy at best. Some branches struggled to get women involved (GPMU 1998b). Many chapels did not have committees on which women could sit. Recommendations from successive GPMU women's conferences made it plain that information rarely appeared to reach women members[7] and policy implementation at local level was minimal, despite EC endorsement, as can be seen from the repeated requests for action on equal pay and opportunities issues at successive GPMU women's conferences.

R2(7) was very positive about the GPMU, in light of her subsequent experiences of post-GPMU mergers. She said:

> As there were few women at national level [the structures] … gave other women a chance to have a profile and level of experience. It wasn't a "pull up the ladder" approach but trying to expand. It also had broad political views [the women's network] … were quite strong for women. The GPMU moved on quickly with women—did everything possible … but failed young members and it was harder for LGBT. But I have to give Tony Dubbins credit for creating the equality policy adviser [EPA] … I agreed the job description with Tony Dubbins. It needed to be a mainstream issue, not an academic discussion. Members must directly input into structures. But the GPMU was ahead of the game, especially with resources and materials—if we had them now….'

Representation at the main national negotiations with the BPIF did improve. The union negotiating panel included a woman EC member (GPMU n.d.) and the EPA. Unfortunately, the Federation's reluctance to discuss, what they termed, 'social issues' (GPMU 1999b) ensured little was achieved within these negotiations during the 1990s of specific benefit to women, beyond a restatement of legal rights (BPIF 2004–2005—see Chap. 7).

External relations were somewhat better. The GPMU was active in the TUC and Labour Party, contrary to its constituent unions' historical traditions. GPR ensured that women were present on all delegations (GPMU 2000a). From 1997 onwards the union also ensured that there were delegations of lay members to all the TUC's equality conferences, although delegations to the TUC women's conference pre-dated the formation of the GPMU.

In 1998 the union again managed to acquire a woman national officer (GPMU 1998c). She remained the union's only woman national officer until the merger with Amicus in November 2004.

In conjunction with the organising strategy that the union was pursuing to try and reverse serious membership losses, the union supported these structural changes by appointing their only woman researcher as the union's EPA (GPMU 1997a). She had a remit to pursue equality for black and ethnic minority members, women, disabled and lesbian, gay, bisexual, transgender and young members; the first time the union had explicitly recognised the totality of diversity issues. Arguably, this shift was under-resourced for tackling such entrenched problems, in common with other unions (for instance, McBride 2001), and she was not a national officer.

The strategy included introducing new policies and initiatives to raise awareness of discrimination among the overwhelmingly white, male membership. The women's committee became far more proactive, rather than simply endorsing reports (GPMU 1997b). The EPA was involved in all areas of the union's activities, and networks were established to encourage the involvement of all disadvantaged and underrepresented groups, underpinned by a proposal to establish a workable equality committee structure (GPMU 1999c). Attempts at forming an electronic network to extend involvement further were unsuccessful (GPMU 2001b).

Much more guidance and many new and updated policies alongside greater visibility in the union's journal raised the women's profile (GPMU 2000b). Fringe meetings were held on racism and homophobia at the union's 1999 conference (GPMU 1999e). Training for women members, lay representatives and full-time officials (excluding national officers) was organised. However, local implementation remained poor, as in other unions (Munro 1999; McBride 2001).

This strategy needed to be successful given that women's membership fell 44 per cent compared to 35 per cent for men between 1991 and 1998 (Roe 2003). It appeared that women's higher profile in the union did bear fruit. Among new recruits, at the highest point, women constituted 28 per cent approximately (GPMU 2004) compared to the 17 per cent of total membership they, by then, represented. Organising and equality initiatives were slowly beginning to change the union's profile.

The law, however, had a limited impact on improving women's representation. In particular, in 2000 the union settled a case of sex discrimination against a member claiming a breach of law by both the company and union combined (EOC 2001). The lessons learned from this case helped to reinforce the need for a greater understanding of equality issues, as misrepresentation was an increasing possibility where union officials failed to see the implications of discrimination in grievances. Branch full-time officials were pressured into attending training that provided equality awareness as well as knowledge of the law (GPMU 2002). But the effect was to refer more individual grievances to lawyers rather than seek industrial solutions, unless employers could be pressurised into a settlement to avoid legal action.

Undoubtedly women's representation increased dramatically during the GPMU's existence, but this was occurring during a period of equally increasing weakness. Men's technical power was beginning to fade, while the political and legal climate, discussed in Chap. 3, further undermined their sexual and collective power. Still their response to women, at a time when work was generally becoming more feminised, was too little, too late (Dawson 2010), and despite women's increasing visibility, the employers were able to outmanoeuvre any bargaining for women. They used their symbolic power to designate equality claims as 'social issues', a subject revisited in Chap. 7. As the previous chapter demonstrated, the GPMU could no longer survive alone, and independent print unionism came to an end in November 2004.

Conclusions

It is clear that from the outset, everyone had an allotted place in craft-based industries, such as print. Women were largely excluded from production work. As the labour process divided, women were allocated the lower-paid jobs, and power resources were mobilised to ensure that they were unlikely to find a route out of this work. The earliest example, in bible printing, demonstrates the degree of women's low pay—but also how disputes over women's low pay could be used as a bulwark against threats to higher-paid jobs too. Therefore, the class battle, between employers mobilising economic and positional power and men utilising sexual, collective and technical power, ensured women were restricted to low-paid jobs. On the odd occasion that women managed to slip these bonds, they lived on borrowed time, as in 1870s Edinburgh, where they lost skilled work as soon as men recovered sufficient sexual, collective and technical power to force them out. Never was there any thought of building solidarity across gender lines once these women were in skilled work, exclusion was the only response. For bookbinders, segregation was the answer, as women came into binderies in great numbers. When segregation into separate women's unions failed to prevent employers segmenting women to access their cheap labour, they were brought into men's unions but still segregated in employment and with limited representation within the union.

In the union, personal power resources, knowing the 'rules of the game', became important in containing women's representation. While these manoeuvres would have been justified as a class response to the war of attrition between workers and employers, they had clearly gendered outcomes. It is hard to accept this justification for retaining such absolute sectionalism because evidence demonstrates women were prepared to be militant whenever possible (McDonald 1904; Bundock 1958), so it would seem greater gendered solidarity might have reaped benefits, especially in the long-term when technology changed elements of the labour process, creating new segments (see Chap. 5).

While the later twentieth century did see greater involvement of women in unions and the NUPB&PW (later SOGAT) did consistently raise issues of women's pay, craft organisation, and its attendant segmentation, tended to stymy potential gains for women. Even the Equal Pay Act had limited effect (see Chap. 7). By the time the GPMU was formed, technology was biting deep into union organisation, women were losing their jobs in considerable numbers, and skill was losing its value in bargaining. The union, arguably, failed strategically because it did not follow the job into new areas, as the industry elided with the communications sector. This sector was more diverse than traditional printing. The potentially game-changing problems occasioned by such technological change and industry restructure are explored further in the next chapter.

Notes

1. The Taff Vale case involved the railway company of that name suing the Amalgamated Society of Railway Servants for losses the company incurred as the result of a dispute. This decision was a huge blow to trade unions as it made them liable for damages whenever they took industrial action (Gardiner and Wenborn 1995: 731).
2. Dilutees were substitute workers who were seen as reducing or weakening the skill content of a job because they were untrained or had little training in that job. They were seen as a way of undermining the superior rates paid to those who had done a full seven-year apprenticeship.
3. Correspondence between company and branch and branch and FoC (shop steward), March 1975.
4. In interview, national officer 1 recalled that gaining the house rate had previously been a TA policy.
5. Researchers at the Polytechnic of North London compiled this report, based on questionnaires and interviews with officers and women members.
6. The best example is UNISON. Initially their completely self-organised groups were not linked to the decision-making processes of the union, although they received union funding. Discussions with UNISON officials suggest that this may now have been addressed.

7. See GPMU women's conference Reports—(1991: 1), (1992a: 8, 10, 12, 13, 15, 18, 21), (1994a: 9–10, 12, 17), (1996c: 21), (1998d: 11, 13, 15, 17), (2000c: 9–10).

References

Avis, F. C. (1971). *The Early Printers' Chapel in England*. London: Avis.

Barron, R. D., & Norris, G. M. (1991 [Originally 1976]). Sexual Divisions and the Dual Labour Market. In D. Leonard, and S. Allen (Eds.) *Sexual Divisions Revisited*. London: Macmillan.

Braverman, H. (1974). *Labor and Monopoly Capital: The Degradation of Work in the Twentieth Century*. New York: Monthly Review Press.

British Printing Industries Federation (BPIF). (1988). *Achieving a Balance: A Guide to Equal Opportunities in the Printing Industry*. London: BPIF Joint Training Council.

Bundock, C. J. (1958). *The National Union of Printing, Bookbinding and Paper Workers*. London: Oxford University Press.

Burchell, B., & Rubery, J. (1994). Divided Women: Labour Market Segmentation and Gender Segregation. In M. E. Scott (Ed.), *Gender Segregation and Social Change*. Oxford: Oxford University Press.

Child, J. (1967). *Industrial Relations in the British Printing Industry*. London: George Allen & Unwin.

Clegg, S. R. (1989). *Frameworks of Power*. London: Sage Publications Ltd.

Cockburn, C. (1983). *Brothers: Male Dominance and Technological Change*. London: Pluto Press.

Colgan, F., & Ledwith, S. (2002). Gender, Diversity and Mobilisation in UK Trade Unions. In F. Colgan & S. Ledwith (Eds.), *Gender, Diversity and Trade Unions: International Perspectives*. London: Routledge.

Connell, R. W. (1983). *Which Way Is Up?* London: Allen and Unwin.

Cunnison, S., & Stageman, J. (1995). *Feminizing the Unions*. Aldershot: Avebury.

Dawson, T. (1993). *A Case Study of Derecognition of the National Graphical Association in the Provincial Newspaper Industry: The Kent Messenger Dispute*. MA Thesis Unpublished, Keele University, Staffs.

Dawson, T. (2010). Out of Sight, Out of Pocket: Women's Invisibility in the British Printing Industry and Its Effect on the Gender Pay Gap. *Historical Studies in Industrial Relations, 29/30*, 61–98.

Dean, B. (2007). *Hot Mettle: SOGAT, Murdoch and Me*. London: Politico's Publishing.
Drake, B. (1920). *Women in Trade Unions*. London: Virago Press.
Duffy, P. (2000). *The Skilled Compositor, 1850–1914: An Aristocrat Among Working Men*. Aldershot: Ashgate Publishing.
Equal Opportunities Commission (EOC). (2001, July 10). *Press Release.*
Game, A., & Pringle, R. (1983). *Gender at Work*. London: Pluto Press.
Gardiner, J., & Wenborn, N. (1995). *The History Today Companion to British History*. London: Collins and Brown Limited.
Gennard, J. (1990). *NGA: A History of the National Graphical Association*. London: Unwin Hyman.
Gennard, J., & Bain, P. (1995). *SOGAT: A History of the Society of Graphical and Allied Trades*. London: Routledge.
Graphical, Paper and Media Union (GPMU). (1990). Instrument of Amalgamation, Second Schedule, p. 45.
Graphical, Paper and Media Union (GPMU). (1991). Women's Conference Report, p. 1 and Appendices A and 6.
Graphical, Paper and Media Union (GPMU). (1992a). Women's Conference Report, 8, 10, 12, 13, 15, 18, 21.
Graphical, Paper and Media Union (GPMU). (1992b). Women's Representation on the Executive Council, Report to the Executive Council.
Graphical, Paper and Media Union (GPMU). (1994a). Women's Conference Report, 9–10, 12, 17.
Graphical, Paper and Media Union (GPMU). (1994b). Branch Officers List.
Graphical, Paper and Media Union (GPMU). (1995). Biennial Delegate Conference Report, 8–9, 91–92, 99–100, 110–112.
Graphical, Paper and Media Union (GPMU). (1996a). Circular No. 123.
Graphical, Paper and Media Union (GPMU). (1996b). Minutes of the Women's Committee, 15th May, p. 1.
Graphical, Paper and Media Union (GPMU). (1996c). Women's Conference Report, p. 21.
Graphical, Paper and Media Union (GPMU). (1997a). Circular No. 16.
Graphical, Paper and Media Union (GPMU). (1997b). Women's Committee, 18th April, 2–3.
Graphical, Paper and Media Union (GPMU). (1998a). Circular No. 107.
Graphical, Paper and Media Union (GPMU). (1998b). Women's Committee Minute 98/7.
Graphical, Paper and Media Union (GPMU). (1998c). Circular No. 259.
Graphical, Paper and Media Union (GPMU). (1998d). Women's Conference Report, pp. 11, 13, 15, 17.

Graphical, Paper and Media Union (GPMU). (1999a). Biennial Delegate Conference Report, p. 85.

Graphical, Paper and Media Union (GPMU). (1999b). Ballot Paper on Terms of Settlement for British Printing Industries Federation National Agreement – Conclusions.

Graphical, Paper and Media Union (GPMU). (1999c). Equality Report, 12.

Graphical, Paper and Media Union (GPMU). (2000a). Rules of the Graphical Paper and Media Union, 50–51.

Graphical, Paper and Media Union (GPMU). (2000b). Equality Handbook.

Graphical, Paper and Media Union (GPMU). (2000c). Women's Conference Report, 9–10.

Graphical, Paper and Media Union (GPMU). (2001a). Circular No. 55.

Graphical, Paper and Media Union (GPMU). (2001b). Internal Correspondence, 7th December.

Graphical, Paper and Media Union (GPMU). (2002). Internal Correspondence, 18th November.

Graphical, Paper and Media Union (GPMU). (2003). Biennial Delegate Conference Report, 10.

Graphical, Paper and Media Union (GPMU). (2004). Report – GPMU Organizing Experience.

Graphical, Paper and Media Union (GPMU). (n.d.). List of Executive Council Committee Members.

Howe, E. (1950). *The Federation of Master Printers 1900–1950*. London: Federation of Master Printers.

Hufton, O. (1995). *The Prospect Before Her: A History of Women in Western Europe, Volume 1 1500–1800*. London: HarperCollins.

Ledwith, S., Hayes, M., Joyce, P., & Gulati, A. (1985). *Women in SOGAT '82: Report of a Research Project into the Role of Women in the Union Carried Out by an Independent Research Team from the Polytechnic of North London*. Essex: SOGAT.

Lown, J. (1983). Not So Much a Factory, More a Form of Patriarchy: Gender and Class During Industrialisation. In E. Gamarnikow et al. (Eds.), *Gender, Class and Work*. London: Heinemann Educational Publishers.

McBride, A. (2001). *Gender Democracy in Trade Unions*. Aldershot: Ashgate.

McDonald, J. R. (1904). *Women in the Printing Trades*. London: P. S. King and Son.

Moran, J. (1964). *NATSOPA Seventy-Five Years: A History of the National Society of Operative Printers and Assistants (1889–1964)*. Oxford: Oxford University Press.

Munro, A. (1999). *Women, Work & Trade Unions*. London: Mansell.

Musson, A. E. (1954). *The Typographical Association*. London: Oxford University Press.

National Graphical Association (NGA). (1976). Biennial Delegate Meeting Report, p. 24.

National Graphical Association (NGA). (1988, *Print* April). Mark Cooper's Seven Stages to Becoming a Man and a Real Printer, pp. 6–7.

National Institute of Industrial Psychology. (1969). Selection and Training of Keyboard Operators in the Printing Industry – Final Report.

Nicholson, J. (1994). *The Influence of the London Women's Branch of the Printing Unions on Women's Employment in the Industry: Demarcation, Inter-Union Strife and Employment Restriction 1950–1960*. Unpublished Dissertation, London College of Printing.

Phelps Brown, H. (1977). *The Inequality of Pay*. Oxford: Oxford University Press.

Reynolds, S. (1989). *Britannica's Typesetters: Women Compositors in Edinburgh*. Edinburgh: Edinburgh University Press.

Roe, A. (2003). *The National Agreement in the Printing Industry – The Exception to Test the Rule*. PhD Thesis Unpublished, Keele University, Staffs.

Society of Graphical and Allied Trades (SOGAT). (1977). Final Court of Appeal into the Format International Dispute, Modern Records Centre, Warwick University (MSS.39/SO/3/9/3).

Society of Graphical and Allied Trades. (1984). Report of the Women's Positive Action Sub-Committee of the NEC to the 1984 Biennial Delegate Council, p. 4.

Society of Graphical and Allied Trades (SOGAT). (1985). *SOGAT Journal Women's Quarterly*, p. 6.

Society of Graphical and Allied Trades. (1986). Biennial Delegate Council Report, p. 7.

Trades Union Congress (TUC). (1955). *Women in the Trade Union Movement*. London: Trades Union Congress.

Typographical Circular. (1941a, May). pp. 82–83.

Typographical Circular. (1941b, September). p. 137.

Walby, S. (1986). *Patriarchy at Work*. London: Polity.

Webb, S. and B. (1913). *Industrial Democracy*. London: Webb, S. and B. for Workers' Educational Association.

Zeitlin, J. (1985). Engineers and Compositors: A Comparison. In R. Harrison & J. Zeitlin (Eds.), *Divisions of Labour: Skilled Workers and Technological Change in Nineteenth Century Britain*. Brighton: Harvester Press.

5

Gender or Skill? The Continuation of Segregated Work

Fox (1974) argues that, historically, craftworkers and professionals were less likely to accept the ideology of management prerogative. They wished to retain a significant degree of discretion in their jobs, and therefore, discretion was a source of power for them. Workers would seek to enhance their discretion in order to get greater control of the job and may even try to constrain management through gaining agreement to rules that limited management's discretion. This scenario is a clear description of class forces in the printing industry and especially the clash of technical and collective power (workers) and economic power (management). However, it will become clear that class impacts on gender relations in ways that partially helped women but also trapped them in segregated jobs within a segmented labour market, which was a key factor in their valuation (Young 1981; Pollert 1983). Even changes in the labour process were insufficient to the dislodge segregation; gender division is constantly reasserted. These processes are investigated below, first through identifying the employment trends and extent of occupational segregation in the industry, then through an examination of job content and, finally, through exposing the barriers to change. It will be argued that work organisation is at the root of women's invisibility in jobs and, consequently, pay structures and that

T. Dawson, *Gender, Class and Power*, https://doi.org/10.1057/978-1-137-58594-3_5

managers and men mobilised power resources to protect their interests. Recent research is used to comment on subsequent developments post-2002.

Occupational Segregation and the Labour Process in General Printing

Hakim (1979) identified the extent of occupational segregation across sectors over seven decades, demonstrating its persistence. In printing it is clear such segregation was embedded from the outset. Braverman (1974) explained the genesis of division in the labour process, and, despite criticisms that his analysis focused too much on transitions from old craft traditions and ignored workers' resistance (Burchell et al. 1994), his analysis has much of value to offer a gendered reading of printing industry processes. Building on the discussion in Chap. 4, women's resistance to these processes is evident, especially in relation to the 'naturalising' of the sex typing of jobs.

Looking back at groundbreaking gender studies in the workplace (West 1982; Pollert 1983; Cockburn 1983), it is clear that the labour process is a key component of inequality regimes. How structures develop and are maintained is just as important as who fills the 'empty places' in them (Hartmann 1981). Chapter 3 recounted the early division of labour that split preparation of material for printing (compositors) from printing (machine minders), demonstrating the separation of conception from execution in its broadest sense. No worker had complete control of the whole process from the late nineteenth century onwards, while women had been inserted as detail workers in the bindery division of labour (Duffy 2000). This process incorporated a slippage of skill definitions for pragmatic reasons (Hacker 1990) when 'crude mechanics', the machine minders with no craft training, were allowed membership of the craft union. Therefore, given the 'natural' departmental split, gendered occupational segregation was, partially, inherent in the general division of labour and exacerbated by the creation of multiple unskilled jobs in the bindery. The social construction of skill (Phillips and Taylor 1980) is clearly visible in this example. Similarly, the designation of women's tacit skills as unskilled (Beechey 1986; Cockburn 1988) cemented division. It has been suggested that sex typing of jobs can only change with radical restructuring (Liff 1986), although re-segregation could emerge (Cockburn 1988; Rubery and Fagan 1995). The enactment of these processes is visible in the discussion that follows. But first trends in employment make clear the changes taking place in the industry at the time the research for this book was undertaken.

Women constituted about 33 per cent of the workforce in the printing sector (Labour Market Trends 2003). Remarkably there was little change over at least 30 years, as in 1973 women constituted 33.1 per cent of the sector (TUC 1973). However, Cockburn indicates that women's employment fell from 38 per cent through the 1920s, 1930s and 1940s to only 30.6 per cent in 1980 (Cockburn 1983: 159–160), suggesting that maybe the recovery in women's proportion of the industry post-1980 was due, in part, to the decline in male employment observed in official statistics over the same period. From 1980 to 1996 men's employment declined steadily by about 41,000 to 186,600, recovering slightly to 206,200 by 1999 before declining again to 190,100 by 2004. Conversely, women's employment rose over a similar period by about 29,000 from 112,500 to 141,100, falling during the recession to 132,900 in 1991, before reaching its high point of 141,800 in 1994. Thereafter there were fluctuations, but, overall, the reductions were drastic with only 92,700 women in the sector by 2004. Almost 50,000 women were removed from the industry in ten years.

Disaggregated statistics reveal the beginnings of a shift between employment in printing and publishing. Figure 5.1 looks at full-time employment and demonstrates the slow decline in men's employment in printing, compared to a steady but slow increase in publishing. For women publishing employment overtook that in printing by 1999 and, although publishing showed a decline in 2003, it was starting to pull away from printing again in 2004. Part-time employment in Fig. 5.2 highlights the low level of men's employment in both sectors, although there is a steeper rise for men in print. The more interesting figure is in women's employment: a steep decline for women in printing and a fairly steady upward trajectory for women in publishing. While there are no

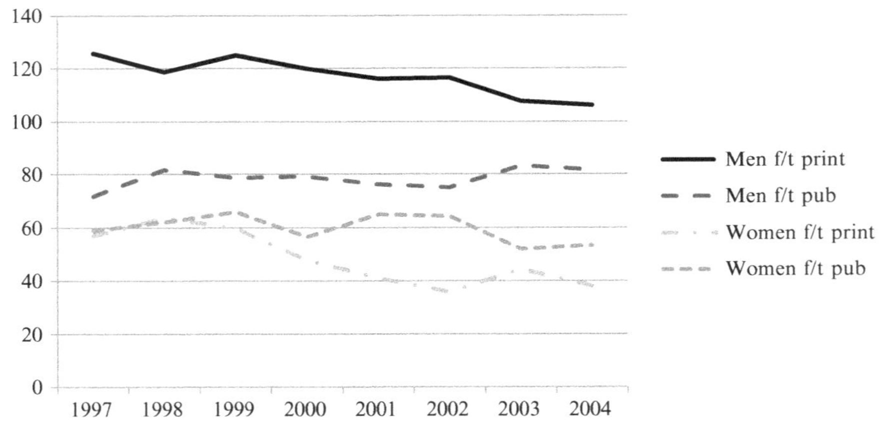

Fig. 5.1 Full-time employment in printing and publishing, 1997–2004 (thousands). Source: Labour Market Trends, 1997–2004

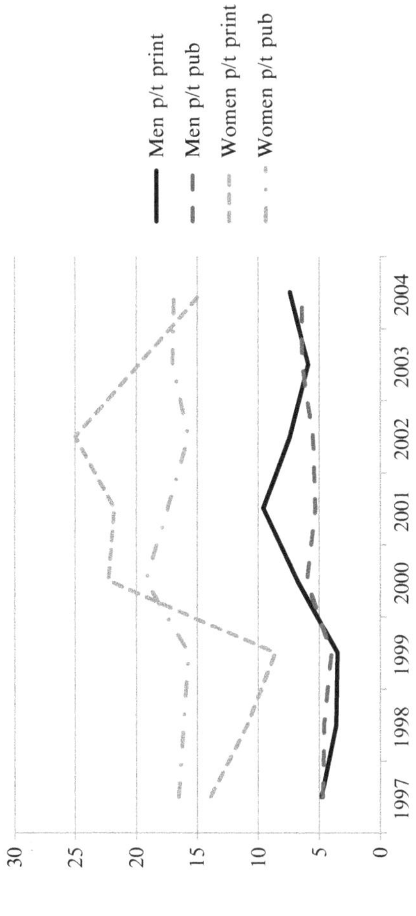

Fig. 5.2 Part-time employment in printing and publishing, 1997–2004 (thousands). Source: Labour Market Trends, 1997–2004

comparable figures available to disaggregate these sectors now, there are statistics on information and communication workers.

Figure 5.3 provides support for the increasing scale of employment in this sector from 2004 when the previous data ended. While they are not entirely comparable, this does seem to indicate the potential for a technology-driven shift from traditional printing from that time. In fact figures in this time series from the 1990s reveal there were around 250,000 women and 500,000 men employed, when recording of employment under this heading first began in 1997.

The employers' federation claimed that some 70 per cent of print sector employment was Class 1, that is, skilled. The Graphical, Paper and Media Union's (GPMU) density in the sector was about 60 per cent at the time of this research in 2002.[1] In the union women accounted for about 17 per cent of membership (GPMU 2002).

Table 5.1 reveals that the definition of skilled work was largely coterminous with the departmental division of labour. The pre-press and printing departments were composed of almost entirely skilled men, and the bindery was split between a number of skilled occupations, almost entirely filled by men and a mass of occupations designated unskilled, or possibly semi-skilled, which were filled by women.

Table 5.1 clearly demonstrated the continued concentration of women in the finishing area (52.8 per cent). Men occupied a similar proportion (i.e., 53.8 per cent) in pre-press and printing where there were small numbers of women. The 24 per cent of men in finishing would mostly be in the skilled occupations. The 'other' category covered such jobs as office workers, including managers with a print background who kept their union cards. Also included in this group were warehouse and distribution workers, cleaners and labourers. The extent of occupational segregation was even more apparent in Table 5.2.

Table 5.2 reveals that even in the one skilled area that women managed to break into, the pre-press department, they were still very much a minority. In the 1980s women were taken into these jobs because of their typing skills, employers taking advantage of skills acquired in their general education (Craig et al. 1984). The temporary nature of this incursion is discussed below, but these figures confirm that women were not employed in pre-press areas in any great numbers.

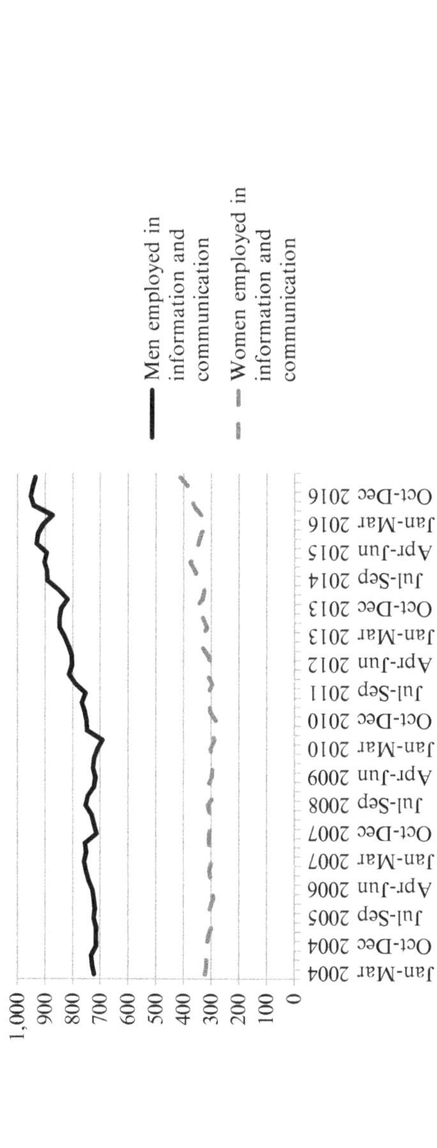

Fig. 5.3 Women and men employed in information and communication sector, 2004–2017 (thousands). Source: ONS (2017)

Table 5.1 Departmental proportions by gender

	Women (%)	Men (%)
Pre-press	6.1	15.3
Printing	4.0	38.5
Finishing/bindery	52.8	23.5
Other	34.2	20.6
Non-response	3.0	2.1
Total	100	100

Source: Survey of GPMU members 2002 (N = 1216)

Table 5.2 Gender proportions by department

	Pre-press (%)	Printing (%)	Finishing (%)
Women	26.3	8.9	66.5
Men	69.3	88.2	30.9
Non-response	4.4	3.0	2.5
Total	100	100	100

Source: Survey of GPMU members 2002 (N = 1216)

It has been observed that men are more likely to work with men, while women are more likely to work in mixed areas (Hakim 1979; Cockburn 1988). Table 5.3 confirmed this pattern in general printing.

Only 2.6 per cent of men worked mostly with women, whereas over 83 per cent worked mainly with men. Conversely for women the split was more even. The figures indicated a slightly higher percentage of women working with an equal number of men and women, compared to working with mostly men. Given that overall women were a smaller percentage of the workforce than men and were generally confined to finishing, this could account for the asymmetry in the figures where 37.1 per cent of women worked with an equal mix of both women and men but only 23.5 per cent of men worked in finishing (Table 5.1). The effects of internal labour markets (ILMs) highlight the nature of segregation and its relationship to labour market segmentation.

Doeringer and Piore provided a model of internal labour markets that highlights the key features of a primary labour market—'high wages, good working conditions, employment stability, chances of advancement' (1985: 165), among others. However, non-craftworkers, especially women, would be expected to have the exact opposite characteristics in

Table 5.3 Predominant gender of colleagues

	Colleagues				
Sex	Non-response (%)	Mostly men (%)	Mostly women (%)	Equal mix of both	Total (%)
Male	1.0	83.4	2.6	13.0	100
Female	1.8	35.0	26.1	37.1	100
Total	1.4	60.0	14.0	24.7	100

Source: Survey of GPMU members 2002 ($N = 1216$)

their secondary labour market model. The more sophisticated variant, posited by Loveridge and Mok (1979), had more salience in printing. This divides into primary external and internal segments and secondary external and internal segments. While most of the primary labour market characteristics were valid in printing, promotion opportunities were generally limited but, in common with findings in other research (Craig et al. 1984; Rubery 1994), still favoured craftsmen in, for example, supervisors' jobs. Also the craftsmen's industry-specific (as opposed to firm-specific) skills were more characteristic of the primary external segment, transferrable to other companies in the industry and supported by common terms, as in the print industry national agreement.

As for women print workers, while generally categorised as low grade, they tended, until recently, to have job stability, limited training and higher wages than unskilled workers might generally expect. This might be categorised as a secondary internal segment (SILM) (Craig et al. 1984: 88). Industry restructuring upset this model, removing women's and semi-skilled men's jobs in large numbers. Anecdotal evidence suggested that print managers' traditional paternalism was changing, compromising 'the credibility of their [existing] labour market policy' (Rubery 1994: 48) and possibly leading to the attempt to create new divisions, for instance, between No. 1 machine minders and all other remaining craftworkers. R2(6), R2(7) and R2(8) all confirmed the reduction in No. 1 positions and the removal of assistants. However, R2(8), in conversation in 2017, had noticed that some chapels were beginning to get the assistants back and skills' shortages were forcing a rethink on training, with a new apprenticeship scheme expected.

The previous willingness to pay women print workers higher rates than unskilled women generally received may suggest two explanations for the SILM—that women were not quite as unskilled as suggested and that the national agreement was effective in providing a higher (although sticky) floor for women's wage rates. It is also clear that while men continued to dominate the skilled occupations, they also tended to dominate the ranks of union negotiators because this formed the recruitment pool from which negotiators traditionally came (Hyman 1994). As a result only limited women's experience has come to the bargaining table, then or since (see Chap. 7).

Research has identified how pay structures may legitimise systems of work organisation and social stratification (Rubery 1997) and how this links to local and national bargaining (Craig et al. 1984). In the case of the national agreement in print, its role in allocating workers to both job roles and grades provides a direct link with pay differentials (ibid.). It was also the benchmark agreement even for non-union firms. Material benefits mainly accrued to men in this structure, as Chap. 7 establishes, so there was little incentive for male negotiators to change them.

Turning to examine the employers, there was little chance of a challenge to the existing job structures from this quarter. Employer 3 registered the only concern with discrimination, but this was more a concern about the reaction of office-based women rather than production women. She acknowledged that the workforce split in her company was somewhere between 25–30 per cent women to 70–75 per cent men on the production shopfloor but not as stark as this in the offices. On the production shopfloor of this company, there was only one MoC (female shop steward). There was no indication of a serious engagement with occupational segregation in the company, but as this manager was the only woman at board level, she may have felt constrained by 'the structural fact of men's supremacy' (Connell 1983: 44) and fallen into [the traditional] line.

Employer 2, from the smallest company, revealed an even lower ratio of men to women than the multinational, with only two women out of 12 production employees (17 per cent). Both these women worked in finishing, along with one man, although some finishing work was also contracted out. Changes in technology at this company had reduced production staff by a third, resulting in work intensification for those who remained.

Employer 5 said that there were very few women in either of their companies. He was keen to stress that one of the few women there had progressed from traditional handwork, over five years, to lead a new venture the company had launched. In the sister company, a man did the setting for the three women employed there, and it was claimed that this was the women's own choice. Employer 4, from a small company, had a more even spread of men and women. This firm employed seven men and six women on production. The women all worked part-time. In the offices there were four men and four women.

Placing this in context, employer 1 said that in his experience 'statistically the level of women employed has not really changed; the proportions in the print room and the binderies haven't changed'. Instead he felt that 'if the industry was taking on a lot of people the pressure would be there [but] the downside is extensive rationalisation, the industry is contracting no doubt about that'. However, this did not address existing job segregation, which is discussed below.

Job Content

The jobs undertaken by member interviewees were important in interpreting some of their comments. Male interviewees tended to work in pre-press or printing. About one-third of male interviewees were employed in finishing, but only one of these was an unskilled worker, a labourer. The women mainly worked in unskilled finishing jobs, with only four working as typesetters, craftwork in pre-press.

Many researchers have identified the heavy and/or dirty nature of work as a reason for women's exclusion from certain jobs (Game and Pringle 1983; Burchell and Rubery 1994; Vogler 1994; Bradley 1999; Munro 1999). This feature has been used to explain women's historical absence from craft occupations. In this respect, employer 3 was clear that strength could no longer be seen as a barrier and so removed some scope for discrimination. But employer 4 still thought that a woman would not want to load a machine eight hours a day, while employer 1 thought these traditional barriers were still in existence. Almost all the union officials thought the heavy and/or dirty nature of the industry had been a barrier in the past, and some of them still thought that pressrooms could be off-putting for those reasons.

In contrast, officer 3 observed that 'women were always pack horses; men have machinery for moving reels', a view supported by other interview evidence. Worker 14 said that the men were supposed to help with lifting but:

> there can be ten items [for feeding] so the stock gets piled up far away from the machine so you have to pick up the boxes and they can weigh up to 15/20 kilos. We should have someone to lift but with ten or twelve machines running it's not possible so you do your own or get shouted at for downtime.

Worker 29 reported that they had just got four women into Class 1. But she felt that some of the Class 1 work would be too heavy, and she remarked 'the old timers would not lift a box for you, the new ones are better'. Presumably these women would be forced to do the lifting or vacate the positions.

Men may see things differently, as one indicated,

> I felt in principle women should have equal pay. However, in my experience, there are sometimes circumstances in which men are expected to carry out some of the heavier tasks but do not benefit with extra pay to reflect this.

So, while skill is often cited as the justification for higher pay, sometimes, in reality, heavy lifting becomes the key difference (Cockburn 1983). The above demonstrates how women's use of strength has frequently (conveniently?) been ignored (Walby 1990).

Another important factor that divided men and women, even if both operated machines, was machine setting. The following tables indicate who ran machines and, more crucially, who set them.

Table 5.4 showed that while many women ran machines (53.3 per cent), far more men did so (71.2 per cent). However, this still meant that over half of the women were machine operators. Table 5.5 indicates the degree to which women and men set machines.

Almost twice as many men as women set machines, which had significant implications for positions in both the job and pay structures. One survey question asked people to name the machines they operated. The intention was to ascertain the relative complexity of the machines operated by men and women as a factor in occupational segregation and pay. What the question could not do was ascertain the frequency of these tasks or the size of the machines. However, the training would be the same whether the printer was running a four-colour press or an eight-colour press.[2] It was skill acquisition for large, expensive machines that primarily accounted for pay levels, although pay increased the bigger the machine. For instance, worker 8 said, 'I run everything from a four-colour to a single colour'. At this stage flexibility among men in finishing was more apparent than in the pressroom. One survey example included: 'Opticut high speed sheeter, collators, encoders, guillotine, stitcher, driller, hooklocker, detachers'.

Table 5.4 Percentages of workers operating machines, by sex

	Machines			
Sex	Non-response (%)	Yes (%)	No (%)	Total (%)
Male	1.6	71.2	27.2	100
Female	1.8	53.3	44.9	100
Total	1.9	62.4	35.7	100

Source: Survey of GPMU members 2002 (N = 1216)

Table 5.5 Percentages of those running machines who also set them, by sex

	Setting			
Sex	Non-response (%)	Yes (%)	No (%)	Total (%)
Male	15.3	68.9	15.8	100
Female	29.4	36.6	33.9	100
Total	22.2	53.2	24.5	100

Source: Survey of GPMU members 2002 (N = 1216)

The women's response to this question confirmed their long-standing flexibility and expanded on the question about whether they set machines. Examples included, 'Muller/folder/stitcher/drill/collating/shrinkwrap—but I do not set the Muller or folding machines'. Another example suggested an element of team working—'3 of us look after strappers + film wrap units + palletiser. While operating them we set them as parts are worn', or 'Wilstead, Bielomaatic (but these are the only ones I do not set), enveloping machines, stitching machines', or again, 'drill, gatherer stitcher (saddle), perfect binder, wire-o-punch, Muller (trim, stitch) coater (only set this one)'. A woman delegate forcefully made the point about women's flexibility at an earlier SOGAT conference:

> I have with me a list of women's work graded class 3, the lowest. There are forty-nine different operations, and women in the course of a week's work could be called on to do every one of them … A great many of our firms employ only two or three women, therefore, interchangeability is a must. So it is not unusual to find yourself doing a dozen or more different job operations in the course of a day's work. (SOGAT BDC 1976: 398)

It seemed that women may set the odd one or two machines, but as they were not regarded as complex, it did not confer skilled status. As noted elsewhere (Craig et al. 1984; Beechey and Perkins 1987), informal setting may be undertaken without recompense, for instance, to keep old equipment running. That men and women ran different machines could exacerbate the gender pay gap (Rubery 1994). But of equal concern was the potential for women's jobs to become men's jobs (Hacker 1990; Bradley 1999), such as in the above examples, where both operated collators, a job formerly regarded as 'women's work'. Such developments could partially account for the disposal of women workers in large numbers.

Survey evidence indicated that for male printers flexibility seemed to consist of running different-sized machines, such as a six-unit press and an eight-unit press, although in a minority of instances, 'if we're not busy we could go and help the girls collating but generally we're on our own job' (worker 8). Worker 22 commented that they were now expected to do the jobs that 'the Natties [former NATSOPA members] used to do'.

This is a typically disparaging reference (Cockburn 1983: 130) to members of one of the GPMU's constituent unions. Male NATSOPA members were generally employed as machine assistants. This provided an example of where demarcation had disappeared, with the assistants' jobs often being telescoped into the printers' jobs through technological developments, an 'economics of scope' (Hacker 1990: 183). Such processes could expand to male employment too, narrowing the highest pay to No. 1 machine minders, who have overall responsibility for the printing presses, a process R2(7) and R2(8) had witnessed.

Examples of flexible working among the male survey respondents included, 'I not only do finishing but printing' or 'van driver, forklift truck driver, packing, palleting, get paper up for machine, assist on machine'. While examples from women extended beyond even their normal extensive flexibility. These included unusual combinations such as 'I do finishing, pre-press and office work'. Such examples are noteworthy because they encompass cross-departmental flexibility, something rarely encountered, although not unheard of (Craig et al. 1984). This may have either reflected the way small companies were facing increasing competition and/or that typesetting was often done by the customer, requiring only formatting to be done in the print shop. The only other type of flexibility tended to reflect women's traditional, domestic role in undertaking some cleaning and tidying of the workspaces and, for worker 2, 'doing the garden'. It should be noted that far fewer women responded to this question than men, perhaps because there was not much more scope for flexibility in their jobs. So flexibility could encompass a range of duties within the traditional departmental demarcation, but it could also break this down, requiring duties to be performed across departmental lines, albeit generally without upsetting traditional occupational segregation.

Work intensification was evident but with no lessening of the segmentation between men, while craftworkers maintained their sense of superiority (Cockburn 1983). Gendered segmentation was also apparent in relation to who set machines, thereby affecting access to extra payments. Importantly it also revealed slippage between men's formal duties and women's informal (setting) activity; so the process was gendered in relation to the recognition of actual job content. Where demarcation had slipped across departmental boundaries, there was little evidence of women acquiring access to men's jobs, but some evidence that men were incorporating women's work into their own job roles. The tendency, then, was for 'job enlargement rather than multi-skilling' (Dickens 1997: 284), and in women's jobs, this rarely received financial compensation. Job loss resulted from this but differentially, as it affected women generally and male machine assistants in particular. So Class 1 occupations were protected, but Classes 2 and 3 became expendable. Therefore desegregation of the labour process did not remove barriers to segregation but caused job loss that reinforced it, as Cockburn noted (Cockburn 1988). Segmentation was maintained but was adapted to improve competitiveness. It also appears that, subsequently, the typical shift from craft to technician status (Liff 1986) was completed (R2(7) and R2(8)).

While all the employers interviewed had invested in new technology, theoretically removing barriers to Class 1 jobs, women had still not entered craft occupations. Employer 3 explained that they had computer-to-plate (CTP) technology, which was 'a boon to women', but they required few people on it so no new jobs were created.

According to all the trade union officials interviewed, technology had changed at an ever-increasing rate over the last 25 years. Officer 11 explained that women had made temporary inroads into craft sections in general printing and had got the rate for the job. He said,

> [k]eyboard operators in the early 80s were some of the highest paid in the industry, including women … [but] men used to say they couldn't do other bits of the job. Then work came in on disk and so they didn't need the double keystroke.[3] It took four or five years to disappear … A few of the women who had trained up managed to stay in.

Confirming this, worker 23 said that where women had, at some stage, managed to get into skilled jobs, they had not tended to remain there long.

So women entered on the basis of equality but were disparaged because they had not done a full apprenticeship and, for the majority, their success was short-lived. A similar, but more extensive, situation occurred in the newspaper industry (see Chap. 5). By the 1990s many of these jobs had disappeared. New technology broke down the barriers of occupational segregation and then removed the new jobs by which time rates of pay had tumbled, a situation not dissimilar to that in engineering in the US where women displaced men, before themselves being removed (Hacker 1990).

Overall employment in origination (pre-press) had dropped. So while the majority of full-time officials saw this area of printing as much more conducive to women's employment, there were far fewer jobs available. However, officers 9 and 13 believed that there was a gender split of at least 50/50 in the remaining jobs for women, something not apparent in the sample for this survey, although it may have been more relevant in newspapers.

The effect of men rarely working alongside women was reinforced by job losses among women. In worker 10's company, the finishing department got sold off even though it still operated in the unit next door and so there were no women. But work organisation still played a key role. For instance, a couple of women worked in the pressroom but only on forklift trucks (FLT) or sorting, not as printers in worker 25's company. However, this example also demonstrates segregation slippage as, even today, men aim to lay claim to FLT driving. Perhaps this concession was intended to ensure printers' jobs remained solely in men's hands. In some instances, the lack of contact was fostered by departmental organisation. Worker 9 said there were only women in the finishing department. But as he explained, the cutting and creasing operation, normally a bindery job, had been hived off into a separate department. This would exacerbate gender segregation.

In the 30–50 age group, all the men, except one, were Class 1, skilled workers. In only one instance were women working in the pressroom, and they were machine assistants. Worker 6 believed they came up from the finishing department, and he clearly harboured some resentment towards them, saying, 'we've got diversity. There's a news magazine that's full of buzzwords, Americanisms'. While he said he did not object to women in the pressroom, he felt that appointments may just be made to be 'pc' rather than on merit. He clearly felt that one of these two women was getting away with not having to do certain parts of the job—'she's good at work-avoidance techniques'. These appeared to be the only two women he had any contact with. Generally he seemed unhappy about the restructuring that now allowed for 'progression' but resigned to it. This example suggests that younger men may not be any more welcoming than older men towards women, as well as faint rumblings of a possible backlash.

The men mainly reported that women were only employed in the finishing department in all their companies and could not explain why. It often did not appear to occur to them that it could be different. As worker 9 said,

> women are assistants only in finishing but they are not in other areas of the factory. I don't know why. I wouldn't say they would block them. It doesn't make logic—there's not necessarily heavy lifting. I don't think the sex side would come into it. Ten years ago there was a woman trainee in print but she left after being disciplined.

Workers 1 and 2 were aware of women working in the bindery but rarely had contact with them. Worker 1 explained that the women in his company tended to work on the back of the finishing machines, taking the work off the machines; so his experience was of women in traditional roles.

Traditional women's table work had mostly disappeared. In worker 3's company, finishing was fully automated and no women worked on production. The men stated that there were few women still employed in their companies. In addition, although all interviewees had seen changes in technology, the existing workers did not need to be replaced by those with new skills because 'the basics are the same but the tools have changed. What was several trades have now telescoped into one' (worker 23). The work got faster, but with the technology fewer people were needed.

Women interviewees also noted a kind of 'sticky floor' (Davies et al. 2000) keeping them in traditional roles. Worker 26 commented that '[t]he women in finishing, especially packing, do table work and can't get out of it. The finishing machines—guillotine, stapling, folding are all run by men'.

Only four of the women interviewed were in pre-press jobs as typesetters, with one a family appointment, a subject explored more fully below. Worker 30 said all the pre-press used to be women in her company, and this was the reason they had no union organisation in her department, unlike the pressroom. Worker 2 was a printer but only on small machines and also did her own platemaking. Worker 15 explained that in her department 'the blocking machine is Class 1 and the guillotine is Class 1. Very few women are on them. There is one woman on each but the blocking machine woman rarely gets put on it'. Worker 28 reported that there was one female printer in her company. Clearly, even where trained, few women got to work in the better-paid jobs. The next section explores the barriers to changing this situation.

Barriers to Removing Occupational Segregation by Sex

Predictably, employers tended to focus on workers' appropriateness for jobs (Bradley 1986; Liff 1986) and rarely thought job segregation was anything to do with this. Given this inclination, it is perhaps unsurprising that employer interviewees generally believed there were no barriers to women entering craft jobs and struggled to account for its continuation. All the male managers suggested that women did not apply for craft jobs; there was no discrimination. Employer 5 felt that the only progression women wanted was off the shopfloor, even if a job as a machine minder would pay more. Employer 2 thought that historically 'men didn't regard hot metal[4] as a woman's job'. Employer 4 felt that the women tended to be older and set in their ways, wary about new equipment. He also believed that attitudes to occupational segregation were so entrenched he was not sure how to change the mind-set, especially of the 'blokes knocking sixty. Those in their fifties have perhaps seen women do more'.

Questions of appropriateness may also explain, in part, the resilience of stereotypical views about women in binderies. Chapter 7 identifies the likelihood that printing company managers, especially those responsible for production, would have an apprenticeship background. Added to this, the limited experience of working with women makes it less surprising that these managers were at a loss to understand why women never moved beyond the bindery. As McEwen Scott makes clear, regardless which model is used, 'the *mechanism* of segregation remains the market' (McEwen Scott 1994: 158, original italics), but its assumed economic rationality is tainted by deviations from empirical evidence, suggesting stereotypes have greater force (ibid.). These managers failed to comprehend the reasons why women were so reluctant to apply for jobs outside the bindery, a subject discussed in Chap. 6. However, it has been suggested that women may perceive the personal costs of moving out of stereotypical jobs as too high to take the risk (Cockburn 1988). Although there were few examples directly from this research (see Chap. 6), the author had experience of case work that made this situation all too real in the few instances where women sought to move into traditionally male areas of work. R2(7) attested to the importance of the GPMU's Dignity at Work policy in this regard.

Whatever the case, it is clear employers felt no pressure to address this imbalance and, conceivably, feared addressing it would create problems in the disturbance of a homogenous work group. In fact, employer 5 admitted that it could be a relief to employers if they were able to avoid employing women on craft jobs, providing a telling example, which also suggested why so few women seemed prepared to attempt it. Twenty-five years earlier a woman had tried to get into a craft occupation, but the men blocked it, thereby avoiding possible friction on the shopfloor. Women workers also refused to speak to this woman when she finally got a job on a Rotaprint machine.[5] Here the expectations of both female and male workers produced a normative effect, whereby a woman wanting to step out of the traditional role expected of her offended both sexes. Women may often fear their women co-workers' reactions as much as men's (Cockburn 1991). For men such transgressions may offend against their sense of masculinity (Game and Pringle 1983). However, managers want stability and calm; acquiescing to the 'culture of masculinity' (Cunnison and Stageman 1995: 13) in printing may have ensured men would be 'less likely to buck the system' (Game and Pringle 1983: 33).

As stated earlier, evidence suggested that employers were trying to create new divisions within craft occupations, possibly by making No. 1 machine minders the main primary labour segment, reducing benefits to other craft (Class 1) workers. Segmentation was affecting previously inviolate male groups, while women and male non-craftsmen disappeared, shifting the boundaries of segmentation and potentially rendering issues of sex discrimination largely redundant in employers', and possibly workers', eyes. Women had become so invisible through entrenched occupational segregation and the more recent reduction in their numbers that union officials and men no longer saw them as a threat. Employers' lack of inclination to employ women in Class 1 jobs, as confirmed by their comments above, apparently justified this belief. In fact, there was still an impression that men were to blame for blocking women's progression. For example, employer 3 felt that 'there's bigotry amongst men, protecting what they see as male interests'. As McEwen Scott (1994) suggested, resistance to substitution often remained, despite workplace change, making it difficult for employers to use women as a reserve army unless technological change was so radical that it swept away the old labour process (as happened in newspapers), an example that, arguably, exacerbated men's hostility to women in print. Alternatively, as suggested above, it was believed that women did not want craft jobs. Either way, it demonstrated that even for women managers the status quo was hard to alter. Employers were able to absolve themselves of responsibility and blame printers for women's exclusion, as they had for 20 years (PPITB 1982).

Given that women could only act as substitutes when they were 'real competitors for men's jobs' (Liff 1986: 79) and the unions' experience in newspapers, why were male union officials so confident that substitution would not occur in general printing? Generally, they drew a distinction between what happened in newspapers and general print. Two quotes from interviews illustrated their thinking. Officer 1 said that back in the 1960s, when Qwerty keyboards were first introduced into typesetting, women were seen as a threat and 'the union's policy was to take people from the forces who had these skills … and do anything to avoid taking on women'. Where women did enter the trade, they were restricted, 'not allowed to move off keyboards onto design or lay-out etc.'. Even their eventual encroachment into newspapers was limited, 'they were brought in to do just words and spaces, no display' (officer 2), limiting the extent of their ability to replace men. Instead, employers appeared more concerned with taking back the right to manage and cutting costs, rather than replacing men. So even in newspapers, men still retained relative privilege in terms and conditions compared to women, although women retained a more substantial presence, long-term, than in general print pre-press areas. However, employers' economic interests were unlikely to be served by continuing traditional divisions but by segmentation within the craft group. Consequently, men's confidence in their power resources produced a dangerous complacency about gendered job segregation. This may have arisen from either a blind spot, ignoring developing threats, or an instance of 'head in the sand' to avoid having to face the coming shift into the communications sector.

Cockburn explained how managers are driven to utilise women's cheap labour, how men benefit from the gendered identity of their occupations and how this identity must be re-established whenever it is threatened (Cockburn 1988: 33–34). The above example provides evidence of just such a process: re-segregation through employers' and male union members' collusion to insert a cheaper detail worker into the labour process (Braverman 1974) in newspapers. Although this collusion entailed male resistance to employers' economic power, it was aimed at protecting male status, not all workers. This may explain why, in Chap. 7, there is an example of women getting the (basic) rate for the job but not the additional house rates. Male union members' motivation was borne of 'a fear of anybody taking their jobs but it's easy to scapegoat the woman again' (officer 7). Here she alludes to something apparent throughout this account, other men are usually in a much better position to threaten craft status, but hostility is reserved for women's encroachment. Arguably, this reveals men's emotional response to threats to their status, as signified in the mobilisation of power resources to protect gendered sectionalism and the pay hierarchy.

Neither, apparently, was new technology the answer to women's low pay. While generally trade union officials believed that technology, especially digital printing, would make jobs available to women, they also believed that employers never envisaged women as printers. Officer 9 remarked of the employers 'it was not on anyone's radar screen [that women should get skilled jobs]'. Few seemed to think this had changed. Conversely, officer 14 still thought that there was a 'strong male attitude … [in pressrooms]. They don't really want to see women coming in'. He recognised that men still clung to a residual job consciousness (Cole 1939).

While union FTOs also thought that women did not put themselves forward for craft jobs, this was often qualified by recognition of the lack of encouragement for them to do so. Not all were so attuned to these barriers, as officer 11 said '[i]f I had fifty Class 2 men and fifty Class 2 women, all the men would want Class 1 jobs but only some of the women would want it'.

The persistence of, and difficulties in tackling, occupational segregation offered something of an impasse. Shifting employment patterns might provide a way out. Officer 1 noted that in the printing sector, as a whole, employment was growing. This he saw as evidence of the shift from production to white-collar employment. He thought that women would be far more attracted to the new areas of employment, implying that women would like its cleanliness relative to production work. Clearly it also had the potential benefit that Class 1 men would deem this less transgressive.

Underlying these attitudes is the debate about how skill levels are allocated to various jobs. Why is it that women's jobs are automatically regarded as unskilled while many men's jobs are regarded as skilled? The answer, claimed here, is a combination of employers' strategies to divide and rule the workforce and the social construction of skill, in other words class and gender intersectionality. It is an exploration of 'the material realities of the employment relationship and the cultural meanings used to simultaneously understand and shape' (Holgate et al. 2006: 320) work experience that expose the resilience of occupational segregation in the printing industry.

Phillips and Taylor (1980) identified the social construction of skill as more of 'an ideological category imposed on certain types of work by virtue of the sex and power of the workers who perform it' (p. 79) than a rational assessment of ability. Cockburn elaborated on how changes in the labour process take on a gendered meaning. Technological developments in the newspaper industry removed the physical nature of the job and left men feeling emasculated. It removed their control of the job; the political definition of skill that justified male craftworkers' occupational position gave way to managerial initiative (Cockburn 1983). While the role of skill as a bargaining chip had become much less effective, it had not disappeared; men's identity was still closely bound up with skill definitions, both in the pleasure of the work itself and for material benefits (ibid.). Thus, technological change would tend towards re-segregation through the defence of skill definitions (Cockburn 1988), as a form of recognitional domination (Lukes 2005: 120). The shared background of men and managers would make this a more likely outcome, preserving benefits for both: men retained their superior status and material privileges while managers utilised women's cheap labour. This was a useful coincidence of interest (Lown 1983) serving to reinforce women's invisibility. It was possible because of the effect of the sexual, collective and technical power resources men could draw on and the economic and positional power of employers. Personal power allowed these two groups to determine the rules of the game, excluding or segregating women. However, the long-term dangers from this lack of solidarity left all workers vulnerable to new forms of segmentation.

The remaining women in the industry were stuck in the bindery because, as employer 3 explained, they are not in the feeder roles for craft jobs (machine assistants). But employer 1 explained that these were among the jobs in decline anyway. On the one hand, women were trapped by men clinging to the identity politics of skill, despite the dangers. This provided a way for men to manage the contradiction of how to retain their claim to machinery that was becoming progressively less physical and dirty when that was their main explanation for the absence of women in the pressroom. On the other hand, managers were blaming men for job segregation, rather than taking responsibility for how their utilisation of labour market segmentation produced discriminatory outcomes. The lack of concern, or possibly ignorance, about this was inherent in employer 3's claim that their future intention to source the best person for the job should provide opportunities for women, despite the decentralisation in her company, a frequent problem in ensuring equal opportunities (Dickens 1997).

Employer 2 raised another barrier to women entering non-traditional jobs—the effect of long service. More than once he referred to print being 'a long-service industry' where there had been some change but not a lot. He even labelled it 'an incestuous industry'. In his case he pointed out that there had not been a vacancy for seven years in the skilled jobs, so there was little scope for women to move up. Employers 1 and 2 also specifically raised the long-service nature of the industry. As employer 1 pointed out, with the restructuring of the industry requiring fewer people anyway, 'it tends to perpetuate the existing mix [of employees]'. Member interviews also confirmed the long-service nature of the industry.

The extent of low labour turnover was apparent in Table 5.6, taken from survey evidence. This indicated that men had the longest tenure (47 per cent over 25 years) compared to just over a quarter of women (26 per cent). But women were still not short-service employees, given that only just over a quarter (27 per cent) had less than nine years' service. It is probable that the superior benefits of the national agreement, relative to other industries, supported their retention (Craig et al. 1984), but lack of alternatives might also be implicated. The latter issue is explored further in Chap. 6.

Researchers had noticed two related factors of importance for this study: the tendency for different treatment of 'core' and 'periphery' workers and for the polarisation of skills in manufacturing industry (Grimshaw et al. 2001; Gallie 1991). Taken together, they provided scope for a new kind of segmentation in printing. While officer 1 felt that it would be possible for women to get into the pressroom, the instability of the industry and the move towards employing fewer core workers but more agency and part-time staff would militate against it. He had noticed that the industry was polarising into high- and low-skilled jobs. Increasingly men were being employed on low-skilled bindery jobs, and he said, 'the low-skilled problem is now increasingly a low-skilled, male problem rather than a low-skilled, female problem', the latter the result of declining women's traditional manual bindery jobs through technological change. As officer 9 noted, 'even in the bindery where women and men work in

Table 5.6 Length of service, by sex

Years of service	Women (%)	Men (%)
Less than 9	27	12
9–17	24	19
17–25	19	20
Over 25	26	47
Non-response	4	2
Total	100	100

Source: Survey of GPMU members 2002 ($N = 1216$)

closer proximity than in the other departments, women have not moved into craft finishing jobs'. Not only could women not get Class 1 jobs, 'but they can't even become the supervisor of the Class 3 jobs'. The polarisation of skill, and restriction of job ladders, had then already taken place for women in the original division of labour and was now beginning to occur between men, providing scope for new divisions between workers.

Skills' shortages were already apparent in 2002, but revisiting the industry in 2016, interviewees did not think attempts to address the problem would benefit women. In the interim period, finishing departments had often been hived off into separate units or separate companies, increasing women's invisibility. Therefore it was unlikely women would be seen as candidates for the new apprenticeships. R2(8) was still unable to identify any women printers in the industry.

Given the reported lack of friction between men and women on the shopfloor, the male employers had not perceived any problems with occupational segregation. Employer 5 had always perceived good working relations between women and men. He felt that having women around kept 'the more boisterous types calmer'. In fact he remembered the 'apprentices were mothered by the ladies', who brought them cakes. Employer 4 said that men and women got on well, although he acknowledged that some men may think of themselves as better than the women. Acker and Van Houten (1992) suggest that lack of friction is common where men control female groups, and it may also indicate the use of personal or sexual power resources by women as a survival strategy for 'getting on' with men, where resistance is too costly and they receive some material benefit (Lukes 2005).

Trade union officials generally detected no friction either. This was an understandable belief given that women were not perceived as a threat to men's jobs. However, women may also cling to their gendered workspaces as a means of security and to render work bearable (Greene et al. 2002). Officer 12 felt that women did not have the confidence to move into unfamiliar areas, especially where they would be isolated from other women. He felt there might be a bit of 'mickey-taking'. Officer 4 agreed that moving individual women into male-dominated areas was unlikely to prove successful, concurring with other research on changing the sex composition of jobs (Craig et al. 1984; Cockburn 1988). Furthermore, the lack of friction obscured the lack of solidarity, as officer 2 said, 'if you suggest that men use their authority as skilled to push for [equality] it won't happen—they won't do it for men, never mind women'. He noted that the men would not even fight for the casuals (mostly male workers). Conversely, officer 3 said 'there's not enough active support to make a breakthrough; it's not good enough to say that we tried 12 years ago and failed'.

Few women felt able to challenge for craft jobs but comments from members potentially accounted for the lack of overt friction. For instance, worker 2 often felt ignored and thought this was because of her gender. She also felt that the other printer, an older man, deliberately restricted her access to knowledge about different machines. Worker 18 explained 'there's a handful of men think it's beneath them to pick up a brush and clean their areas. It all gets left to the women … there's not a week goes by that we don't have this argument'. Despite its trivial nature, the men's response is not surprising given the symbolic significance of skivvying as denoting unskilled work (Cockburn 1983).

It may be low level, but it does appear that some kind of women's resistance is always occurring (Lukes 2005). However, gendered processes, promoting or limiting militancy (McEwen Scott 1994), appear to intersect with occupational segregation in printing, rendering women's industrial militancy largely unnecessary. Given that disputes tended to be departmental, it would only be when a bindery dispute occurred that women might be called on to act in solidarity. It was the strength of the skilled workers' collective power that would lead to challenging employers; lack of technical power resources made women invisible through much of this process. As such, it was less likely that women would provoke their own disputes or that skilled men would rely on their help in pressroom disputes, although it is possible that skilled men in the bindery might have called on them. Gendered solidarity was probably not seen as crucial.

Finally, the father-to-son route into the industry had largely disappeared, and according to current trade union officials and members, few of whom entered the industry through this route. However, there was a residual 'family effect'. According to officer 4, male craftworkers would commonly obtain jobs for their wives or partners in the bindery, thereby reinforcing the status roles they each occupied in the home, especially that of the main breadwinner. Thus the ideology of family hierarchy could still affect paid work (Delphy and Leonard 1994; Beechey 1986). The 'family effect' was apparent among the majority of companies represented by member interviews, although industry restructuring was starting to reduce the tendency of family members working in the same company. Relocation, in particular, tended to break the chain of 'same company' employment. For instance, worker 19 said, 'when I started, my dad, brother and sister-in-law were there. This has stopped because the company finds it too disruptive if a number of people have to have time off for funerals, holidays and so on'. This had prevented her getting employment for her sister. However, the tendency for emotional attachment to the industry was noted 'once in, they are addicted to it—it's in your blood' (employer 2). Interviews with members, especially those with older women, implied that this was not restricted to men exclusively.

Apart from hoping a younger generation would change attitudes, there was little identification of anything that might change occupational segregation. Certainly, by 2014–2017, interviews confirmed that employment in the sector had fallen drastically, especially for women, reducing pressure for changes to occupational segregation. The growth of agency and casual, often migrant, labour had provided at least as much work for men as women in traditional women's areas (R2(6), R2(7), R2(8)), although this differed by sector of the industry—women were becoming literally invisible.

Conclusions

The core aim of this chapter was to examine how gendered occupational segregation was maintained despite industry restructuring and technological developments. The main factor appeared to be that, although there was some change to the labour process, this was either within existing occupational segregation boundaries or it was simply a case of mechanising lower-paid workers' jobs out of existence, especially women's jobs. The largely forlorn hope (according to the later research) was that either the shift to more white-collar work or generational shifts would break down gendered demarcation. Instead men, albeit a much smaller number, have been able to maintain their superior positions within the industry, while women lost jobs.

During the period of the original research, there was a suggestion that print gets 'in your blood'. This emotional attachment to the industry had both age and gender effects. Men generally held this view, while it was older women—those most likely to have been learners—who held such attachment. This contradicts Gordon's (1972) claim that privileged workers would be more likely to be job conscious. For men, it could be argued that such job consciousness was a key aspect of their identity and motivated their mobilisation of sexual, collective and technical power against perceived threats to their status and pay. Their early access to these power resources was decisive because women never attained such levels of technical power.

Men's position and identity still required maintenance, however. This led to both resistance to and compliance with employers' objectives. Traditionally, collective bargaining and the national agreement had linked occupational segregation and pay, making it virtually impossible to improve pay without the ability to move up to Class 1. Skill was the basis of men's position and identity, and they became ever more desperate to maintain skill as a bargaining chip, even as it eroded before their eyes and led them to place more reliance on physical power as the dividing line between them and women workers, wilfully ignoring women's heavy lifting (Walby 1986). Many processes came into play in seeking to maintain men's position, and employers were happy to play along for the sake of industrial peace until it no longer paid. In such circumstances, it is hardly surprising that union officials would be more likely to identify men's interests as union interests (Munro 1999; and see Chap. 7).

Among these processes was the use of symbolic power to define which jobs were skilled, a recognised means of legitimating authority (Barrett 1980; Phillips and Taylor 1980). Employers benefited from women's cheap labour as a result. By the period under discussion, such valuations of women's skills were embedded and naturalised to the extent that many employers, who tended to come from time-served backgrounds on the shopfloor, could not imagine an alternative. Because of this employers did not look to women, with existing experience in the industry, to relieve skills' shortages, and such flexibility as they accessed was, in common with other sectors (Dickens 1997), mostly within gendered occupational boundaries.

However, more recently employers were struggling to maintain profit margins and needed to rethink their approach in an industry where 70 per cent of the workers were in skilled grades. They moved cautiously, and the maintenance of gendered boundaries would have been reassuring to male workers. The experience of the newspaper sector is instructive. Here compositors were wiped out and replaced by women with pre-acquired skills in typing; wages fell. This likely increased hostility to women moving into skilled jobs. But men seemed to feel that their ability to exclude or contain women in general print meant they were safe. They ignored the segmentation occurring within Class 1, as employers used their economic, positional and symbolic power resources to redefine who got the extra payments and to reduce employment. Therefore, the polarisation in skills arrived in a two-pronged assault: from casual and agency labour replacing the lower skilled, especially women, and from redundancies among the skilled, as well as the re-definition of access to the top pay rates through segmentation within men's jobs.

Women's resistance was generally low level (but see Chap. 6). Instead they either appeared resigned to their position or actively sought to find coping mechanisms. The latter is visible through the example of sexual power being used to look after the men (bringing them cakes). While Chap. 6 discusses women's attempts to break occupational boundaries, this chapter highlights how doing so is recognised as coming with potential serious personal costs (Cockburn 1988). This suggests women adopted a form of rational, tactical compliance (Lukes 2005).

Challenging discrimination has more recently brought the accusation that women are indulging in identity politics at the expense of class solidarity (see, e.g., Fraser 2008). There is an irony to this given the clear exposition of how salient male identity has been, and continues to be, in printing. Arguably the sectionalism of male craft unionism has been far more effective at undermining solidarity, given the shifts in employment that have occurred in this sector. Furthermore, it may have prevented the union from achieving its traditional policy of following the job into new sectors as technology changed. It is argued here that men have been guilty of playing into employers' hands by failing to leave behind increasingly inappropriate tactics to sustain their relative privilege in status and pay, when building gender solidarity might have reaped longer-term security for workers. This could be quite extreme, such as colluding with employers to ensure women introduced into newspaper typesetting were restricted to 'just words and spaces', re-segregating the job (Cockburn 1988). For the most part, however, the contention here is that men's goals in maintaining status and pay were aligned with employers' desires for cheap labour and industrial harmony: a coincidence of interest (Lown 1983).

Notwithstanding this, it is clear that the ploy failed, with women suffering catastrophic job loss and men suffering either job loss or downgrading, so that an ever-smaller group could access high pay. The union could not survive this assault and was forced into a merger, swallowed up into a conglomerate union, Amicus in 2005. The later research indicates that only the strongest chapels have maintained their position. But employers have not had it all their own way. Continuing and deepening skills' shortages have given way to both a new apprenticeship system and to strong chapels beginning to claw back some of what was lost after the global financial crisis. However, women have largely lost the benefit of the secondary ILM they had benefited from, without progression to higher-paid jobs. While the labour process clearly changed, it was within traditional gendered job segregation boundaries, even though skill was becoming an increasingly threadbare justification for this situation.

This chapter has touched on elements of the inequality regime at work in the printing industry. The next chapter breaks down these practices and processes to identify their contribution to entrenching occupational segregation and the gender pay gap, before discussing trade union responses to them.

Notes

1. This figure was established using the GPMU membership data, BPIF data and official statistics.
2. This was especially true for the significant number of men who were apprenticeship-trained. However, the later research seemed to support the move from industry-specific to firm-specific skills today.

3. Double keystroking referred to the process whereby customers produced material on disk and typesetters re-typed it rather than just formatting it as a way of preserving jobs.
4. See Glossary.
5. This was one of the machines that was used in offices and which skilled printers failed to hold on to, although they belatedly tried to get them categorised as printers work by the TUC and failed.

References

Acker, J., & Van Houten, D. R. (1992). Differential Recruitment and Control: The Sex Structuring of Organizations. In A. J. Mills & P. Tancred (Eds.), *Gendering Organizational Analysis*. London: Sage.

Barrett, M. (1980). *Women's Oppression Today*. London: Verso.

Beechey, V. (1986). Women and Employment in Contemporary Britain. In *Women in Britain Today*. Milton Keynes: Open University Press.

Beechey, V., & Perkins, T. (1987). *A Matter of Hours: Women, Part-Time Work and the Labour Market*. Cambridge: Polity Press.

Bradley, H. (1986). Technological Change, Management Strategies, and the Development of Gender-Based Job Segregation in the Labour Process. In D. Knights & H. Willmott (Eds.), *Gender and the Labour Process*. Aldershot: Gower.

Bradley, H. (1999). *Gender and Power in the Workplace*. Basingstoke: Macmillan Press.

Braverman, H. (1974). *Labor and Monopoly Capital: The Degradation of Work in the Twentieth Century*. New York: Monthly Review Press.

Burchell, B., & Rubery, J. (1994). Divided Women: Labour Market Segmentation and Gender Segregation. In M. E. Scott (Ed.), *Gender Segregation and Social Change*. Oxford: Oxford University Press.

Burchell, B., Elliott, J., Rubery, J., & Wilkinson, F. (1994). Management and Employee Perceptions of Skill. In R. Penn, M. Rose, & J. Rubery (Eds.), *Skill and Occupational Change*. Oxford: Oxford University Press.

Cockburn, C. (1983). *Brothers: Male Dominance and Technological Change*. London: Pluto Press.

Cockburn, C. (1988). The Gendering of Jobs: Workplace Relations and the Reproduction of Sex Segregation. In S. Walby (Ed.), *Gender Segregation at Work*. Oxford: Oxford University Press.

Cockburn, C. (1991). *In the Way of Women*. Hampshire and London: Macmillan Press.

Cole, G. (1939). *British Trade Unions Today*. London: Gollancz.

Connell, R. W. (1983). *Which Way Is Up?* London: Allen and Unwin.

Craig, C., Garnsey, E., & Rubery, J. (1984). *Payment Structures and Smaller Firms: Women's Employment in Segmented Labour Markets*. Research Paper No. 48. London: Department of Employment.

Cunnison, S., & Stageman, J. (1995). *Feminizing the Unions*. Aldershot: Avebury.

Davies, H., Joshi, H., Rake, K., & Alami, R. (2000). In K. Rake (Ed.), *Women's Incomes Over the Lifetime*. London: The Stationery Office.

Delphy, C., & Leonard, D. (1994). Class Analysis, Gender Analysis and the Family. In R. Crompton & M. Mann (Eds.), *Gender and Stratification*. Cambridge: Polity Press.

Dickens, L. (1997). Gender, Race and Employment Equality in Britain: Inadequate Strategies and the Role of Industrial Relations Actors. *Industrial Relations Journal, 28*(4), 282–291.

Doeringer, P., & Piore, M. (1985). *Internal Labor Markets and Manpower Analysis* (2nd ed.). Cambridge, MA: D.C. Heath & Co.

Duffy, P. (2000). *The Skilled Compositor, 1850–1914: An Aristocrat Among Working Men*. Aldershot: Ashgate Publishing.

Fox, A. (1974). *Beyond Contract: Power, Work and Trust Relations*. London: Faber & Faber.

Fraser, N. (2008). From Redistribution to Recognition: Dilemmas of Justice in a "Postsocialist" Age. In K. Olson (Ed.), *Adding Insult to Injury: Nancy Fraser Debates Her Critics*. London: Verso.

Gallie, D. (1991). Patterns of Skill Change: Upskilling, Deskilling or the Polarization of Skills? *Work, Employment and Society, 5*(3), 319–351.

Game, A., & Pringle, R. (1983). *Gender at Work*. London: Pluto Press.

Gordon, D. (1972). *Theories of Poverty and Underemployment*. Cambridge, MA: D.C. Heath.

Graphical, Paper and Media Union. (2002). GPMU Women's Committee Report to the GPMU Women's Conference.

Greene, A., Ackers, P., & Black, J. (2002). Going Against the Grain: Perspectives on Gendered Occupational Identity and Resistance to the Breakdown of Occupational Segregation in Two Manufacturing Firms. *Gender, Work and Organisation, 9*(3), 266–285.

Grimshaw, D., Ward, K. G., Rubery, J., & Beynon, H. (2001). Organisations and the Transformation of the Internal Labour Market. *Work, Employment and Society, 15*(1), 25–54.

Hacker, S. L. (1990). *Doing It the Hard Way*. London: Unwin Hyman.

Hakim, C. (1979). *Occupational Segregation: A Comparative Study of the Degree and Pattern of the Differentiation Between Men and Women's Work in Britain, the United State and Other Countries*. Research Paper No. 9. London: Department of Employment.

Hartmann, H. (1981). The Unhappy Marriage of Marxism and Feminism: Towards a More Progressive Union. In L. Sargent (Ed.), *Women and Revolution: A Discussion of the Unhappy Marriage of Marxism and Feminism*. Boston: South End Press.

Holgate, J., Hebson, G., & McBride, A. (2006). Why Gender and 'Difference' Matters: A Critical Appraisal of Industrial Relations Research. *Industrial Relations Journal, 37*(4), 310–328.

Hyman, R. (1994). Changing Trade Union Identities and Strategies. In R. Hyman & A. Ferner (Eds.), *New Frontiers in European Industrial Relations*. Oxford: Blackwell.

Labour Market Trends. (2003). Vol. III, No. 1.

Liff, S. (1986). Technical Change and Occupational Sex-Typing. In D. Knights & H. Willmott (Eds.), *Gender and the Labour Process*. Aldershot: Gower.

Loveridge, R., & Mok, A. L. (1979). *Theories of Labour Market Segmentation*. London: Kluwer Academic.

Lown, J. (1983). Not So Much a Factory, More a Form of Patriarchy: Gender and Class During Industrialisation. In E. Gamarnikow et al. (Eds.), *Gender, Class and Work*. London: Heinemann Educational Publishers.

Lukes, S. (2005). *Power: A Radical View*. London: Palgrave Macmillan.

McEwen Scott, A. (1994). Industrialization, Gender Segregation and Stratification Theory. In R. Crompton & M. Mann (Eds.), *Gender and Stratification*. Cambridge: Polity Press.

Munro, A. (1999). *Women, Work & Trade Unions*. London: Mansell.

Office for National Statistics (ONS). (2017). Table 13 Employment by Industry.

Phillips, A., & Taylor, B. (1980). Sex and Skill: Notes Towards a Feminist Economics. *Feminist Review, 6*, 79–88.

Pollert, A. (1983). Women, Gender Relations and Wage Labour. In E. Gamarnikow et al. (Eds.), *Gender, Class and Work*. London: Heinemann Education Publications.

Printing and Publishing Industry Training Board (PPITB). (1982). *Equalising Job Opportunities for Women in Printing and Publishing: An Attitude Survey Report*. London: Printing and Publishing Industry Training Board.

Rubery, J. (1994). Internal and External Labour Markets: Towards an Integrated Analysis. In J. Rubery & F. Wilkinson (Eds.), *Employer Strategy and the Labour Market*. Oxford: Oxford University Press.

Rubery, J. (1997). Wages and the Labour Market. *British Journal of Industrial Relations, 35*(3), 337–362.

Rubery, J., & Fagan, C. (1995). Gender Segregation in Societal Context. *Work, Employment and Society, 9*(2), 213–240.

Society of Graphical and Allied Trades (SOGAT). (1976). Biennial Delegate Council Report, p. 398.

Trades Union Congress (TUC). (1973). *Report of a Conference of Affiliated Unions to Discuss the Implementation of Equal Pay*. London: Trades Union Congress.

Vogler, C. (1994). Segregation, Sexism and Labour Supply. In A. MacEwen Scott (Ed.), *Gender Segregation and Social Change*. Oxford: Oxford University Press.

Walby, S. (1986). *Patriarchy at Work*. London: Polity.

Walby, S. (1990). *Theorizing Patriarchy*. Oxford: Blackwell.

West, J. (1982). *Work, Women and the Labour Market*. London: Routledge and Kegan Paul.

Young, I. (1981). Beyond the Unhappy Marriage: A Critique of the Dual Systems Theory. In L. Sargent (Ed.), *Women and Revolution: A Discussion of the Unhappy Marriage of Marxism and Feminism*. Boston, MA: South End Press.

6

Challenging Inequality: Employers and Unions

Two of the key parties to the employment relationship are employers and trade unions. Each of these parties has distinct effects on the extent to which discrimination exists in the workplace. Employers have tended to approach these issues through a combination of legal compliance and procedural mechanisms for addressing discrimination: equal opportunities policies, harassment and bullying policies and, more recently, diversity policies (Cockburn 1991; Kirton and Greene 2016). Trade unions, being democratic organisations, have responded increasingly to membership demands for greater inclusion, including committees and conferences intended to provide spaces for different disadvantaged groups to discuss issues specific to them (SERTUC 2012; TUC 2016). In the best cases, these link to the main union decision-making structures and incorporate proportional representation for women. A wider diversity of groups is now recognised, where initially it was mainly women's structures that were developed (see, e.g., Unite). However, there has been only limited success in delivering equality, and the persistence of unequal power resources, it is argued, has reduced the effects of structural change.

T. Dawson, *Gender, Class and Power*, https://doi.org/10.1057/978-1-137-58594-3_6

How this has come about is the subject of this chapter. While the barriers to greater gender equality identified below are not new, they do demonstrate the inadequacy of current approaches to disrupting the accumulation and deployment of power resources. Revisiting earlier industrial relations discussions can throw light on how gender inequality remains so embedded within workplaces and, to a more limited but still significant extent, in trade unions. In particular the debates that arose out of the Donovan Commission report (1968) into trade unions and employers' associations raised the issue of 'proceduralisation'. The Donovan Commission was tasked with bringing order back to the system of industrial relations in the 1960s, and one of its prescriptions was to bring greater formality to workplace bargaining, often through the use of clear procedures to constrain how the parties pursued their interests. The pluralist assumptions behind these recommendations were roundly criticised for their inadequate conception of workplace conflict and for failure to adequately analyse the role of the state in industrial relations (Goldthorpe 1974; Howell 1995). The subsequent shift to an HRM paradigm, as trade unions weakened after 1979, also had recourse to the value of 'procedural justice'. It recognised that clear procedures operated with consistency and transparency could further legitimise management prerogative by gaining consent to management decisions arrived at through procedure, even where the decision was disliked (Bingham 2016).

It is argued in this chapter that a similar process has operated in relation to gender equality in both the workplace and the union and, although procedures have made gender inequality more visible, they have also contained the extent to which transformative change has been possible. These processes have also legitimised existing inequality regimes and prevented the disruption of gendered power resources, thus minimising gains to women at work. The following sections use both survey and interview evidence to demonstrate inequality regimes (Acker 2006) in practice: first in terms of employers' approach to discrimination and equal opportunities and then by considering internal union democracy and the limitations of structural change.

Discriminatory Treatment at Work

The basic asymmetry of power in the employment relationship is well established (Dickens and Hall 1995). The contract's basis in property rights (Selznick 1969) privileges management power over workers. Women have responded through combining their sexual and collective power resources to challenge the discrimination that has become embedded in this relationship. Equal opportunities policies were the employer response (Cockburn 1991), a means of enabling a defence to any claims of discrimination. Since the 1990s a discourse of diversity has encompassed these policies in an HRM approach that individualises discrimination, obscuring the group-based nature of this phenomenon. Employer and union approaches to these workplace issues expose the mechanisms by which discriminatory treatment persists, despite anti-discrimination legislation.

It was clear that the employers interviewed had spared little thought for discrimination. There was no concern to ensure their equal opportunities policies (EOPs) were watertight to secure penalty avoidance (Dickens 2005). For employer 1 the business case was paramount in pursuing diversity, which meant that if diversity management did not serve organisational goals, it was unlikely to be pursued (Noon 2007). However, as part of the Federation's attempts to professionalise management, they were denying indemnity insurance for litigation to members abjuring Federation advice. However, rarely were these discrimination cases, highlighting the limitations of women's legal power. It was, therefore, logical that employers anticipated few problems and saw no need to challenge the status quo. As such, employer 3 revealed that managing diversity was decentralised to plant level, with no specialists employed to ensure compliance. Nevertheless, she felt there was no need for further action beyond ensuring plant managers were kept abreast of current legal developments. No one was driving the diversity agenda and there was no monitoring; all factors linked to a lack of substantive equality and still widespread (van Wanrooy et al. 2011; Kirton and Greene 2016), suggesting the business case for equality had little relevance to this multinational. It also demonstrates the shift from the more social justice-focused equal opportunities policies to the managing diversity approach, whereby organisational goals trumped social justice.

In the SMEs informality and flexibility governed practice. Employer 2 said, 'larger companies have procedures but this can work to their detriment at times because the rules are so specific'. He also bemoaned the implementation 'of so many dictats and legislation … you're so busy printing you can't know about them all'. This company relied on the Federation and on an ex-Federation employee for advice. Employer 4's company relied on informal problem-solving, a factor long recognised as 'particularly likely to facilitate sex discrimination' (Collinson 1988: 1).

While traditionally unions had seen the law as a bargaining tool, in this research there was evidence of a shift towards seeing it as the only means of pursuing equality for at least some officials (see also Chap. 7). This was exacerbated by the lack of knowledge some officials demonstrated about existing union anti-discrimination policies, despite most of these officials being required to attend training on this subject (although some avoided attending), indicating a tendency for officials not to take such policies seriously (Healy et al. 2002). Conversely, all the women officers were well versed in these policies.

Among the male branch officers, officer 11's branch had made considerable efforts to pursue gender equality, although still with significant limitations. While he thought they had had some success in convincing members of the benefits for all of implementing equality policies, he believed that 'the way it's put forward is important. You have to be careful about the negative side. You still … often get the comment "if I was a woman I could claim discrimination"', meaning he felt the need to demonstrate that he was not solely concerned with women and black people's rights. His experience also highlighted the problem of isolated women, for example, where there 'is thirty men and one woman, so why support her? It is a common response.' He also felt that middle-aged women tended to accept their lot, which ignores the constraints on women's militancy (Purcell 1979). Union policing of policy implementation was also lacking (officer 10). While increasing union weakness was likely part of the problem, union survival as the industry morphed into a communications sector should have highlighted the dangers of ignoring women's equality. As officer 1 noted, 'we can't survive and not do that [follow the job into new areas] that's the reality', although he too recognised a lack of commitment on the part of officers to pursue equality policies. However, the growing numbers of women HR managers/directors were not blameless; officers noted that none had raised equality issues, probably indicating the constraints of a male-dominated environment (Connell 1983).

Survey evidence provided some context to gendered workplace relations. Table 6.1 indicates that 33 per cent of employees felt they were well treated, while 56 per cent said the opposite.

There is a 10 per cent gender split in favour of men on receiving good treatment, perhaps because men's collective and technical power resources were providing greater protection than women's. It may also reflect the tendency for closer supervision of women (Bruegel and Perrons 1995).

Table 6.2 delves specifically into perceptions of differential treatment. Some 50 per cent of women agreed with the statement that women received less favourable treatment than men, compared to only 13 per cent of men. Even starker was the difference between men (58 per cent) and women (34 per cent) who strongly disagreed or disagreed with this statement. But the gender segregation of men and women is also apparent in the acknowledgement that 26 per cent of men had no idea if this was true.

Interview evidence initially appeared to suggest few problems in relations between men and women workers, but, on closer inspection, the

Table 6.1 Does management treat all employees well?

	Non-response (%)	Strongly agree (%)	Agree (%)	Disagree (%)	Strongly disagree (%)	Don't know (%)	Total (%)
Male	1.9	3.5	34.6	35.7	18.2	6.0	100
Female	5.7	2.0	25.7	36.5	22.2	7.9	100
Total	3.7	2.8	30.2	35.8	20.4	7.0	100

Source: Survey of GPMU members 2002 (*N* = 1216)

Table 6.2 Women are treated less favourably than men in my company

	Non-response (%)	Strongly agree (%)	Agree (%)	Disagree (%)	Strongly disagree (%)	Don't know (%)	Total (%)
Male	3.7	2.6	10.3	42.5	15.3	25.6	100
Female	5.0	24.0	25.7	26.6	7.6	11.1	100
Total	4.4	12.8	17.8	34.9	11.5	18.5	100

Source: Survey of GPMU members 2002 (*N* = 1216)

picture was not quite so positive. Women claimed that men tended to get better treatment, more opportunities for promotion and avoided work that was 'beneath them'. Men's superior attitude also seemed to grate. Worker 29 commented on past incidents with the printers, 'it has always been them and us. If ever there was any trouble they'd always ask you to back them but they'd never back you', perhaps reflecting the tendency to see women in male domains as 'trespassers' (Pollert 1983).

This picture tends to support the concerns of officer 11, above, in his wariness at being too closely identified with the equality agenda. It also supports the contention that any 'attempt to change behaviour without tackling underlying prejudicial attitudes or underlying reasons for inequality' (Liff 1995: 483) would be doomed to failure, leading to a backlash which would condemn equality initiatives as mere 'special treatment'—both unfair and illegitimate.

An issue of growing concern to the union was sexual harassment. However, this research revealed little direct evidence of its occurrence. But against this the survey suggested a much bigger problem, given that 65 per cent of men and 72 per cent of women indicated that harassment and bullying should be a priority for the union. This evidence rings true to the author whose caseload suggested that women had problems with both managers and men, especially where women encroached into 'men's work'. In one instance, a woman was promoted to work alongside 13 men in a previously all-male department. After two years of daily harassment, she could not cope any longer and came to the union to negotiate her return to the women's area of the print works with some salary protection. Stanko (1988) and Collinson and Collinson (1996) would have been unsurprised at this method of exclusion, which demonstrates the confluence of sexual, collective, technical and physical power resources to enforce gender compliance. R2(7) also highlighted the importance of the Dignity at Work model agreement that the union sought to include in workplace agreements wherever possible, reflecting the frequency with which officers faced sexual harassment complaints.

The next section considers the contribution of particular employment policies to the maintenance of discrimination, particularly the embedding of job segregation. However, the low value placed on women's work also becomes apparent.

Employment Policies

The following sections consider policies on recruitment, promotion, training and working patterns. The processes revealed below capture the resilience of inequality regimes cementing the status quo.

Recruitment and Promotion Processes

Word-of-mouth recruitment policies are heavily implicated in the sex typing of jobs as demonstrated in Table 6.3.

No other method comes close to word of mouth in directing respondents to jobs (over 50 per cent of both sexes), emphasising the importance of social networks (Grieco and Whipp 1986). The level of non-response could reflect that the unions' traditional role as an employment agency (PPITB 1982) had survived better than expected or that there was more internal promotion than expected. Word of mouth was identified in a number of forms in the survey: union recommendation, family/personal recommendation or direct company approaches to respondents. Whichever of these forms was used, it would tend to replicate the gender demarcation of the existing workforce. Three possible reasons are suggested. Management may continue to prefer such methods because they are cheap, identify self-starters and also because they prefer not to risk current group dynamics (Dickens 2005), privileging the acceptability as much as the suitability of candidates (Harris and Morris 1994; Burchell and Rubery 1994).

Table 6.3 Method of obtaining current job

	Current job				
Sex	Non-response (%)	Word of mouth (%)	Advertisement in newspaper (%)	Advertisement on company noticeboard	Total (%)
Male	21.6	52.5	22.7	3.2	100
Female	23.2	51.1	21.7	4.0	100
Total	22.5	51.6	22.3	3.5	100

Source: Survey of GPMU members 2002 (N = 1216)

Employers tended to fall back on the claim that women simply did not apply for skilled jobs. But they also concurred that those making selection decisions tended to be time-served line managers. This is known to replicate the existing 'white [male] monoculture' (Cockburn 1991). Consequently, occupational segregation was unlikely to change if promotions failed to target underrepresented groups. Sixty-nine per cent of men and 76 per cent of women had never sought promotion, although both sexes thought fairer promotion arrangements should be a union priority (75 per cent of women and 70 per cent of men). The lack of promotion opportunities in print may partially explain men's focus on achieving high skill-based pay.

Interviews revealed a number of blockages to women's promotion. Worker 26 recounted the story of a woman printer being so badly treated that she asked to be transferred to finishing where she had stayed for the last eight to ten years. There were few examples of women getting into Class 1 jobs, even fewer remaining in them. Worker 19 commented, 'I might've had a go at it [promotion] but it was a different environment, there's no women there'. Heavy lifting could be a barrier too, demonstrating the resilience of physical power resources. Worker 18 said, 'if you went to a Class 1 to help you [with lifting] it would be "why should I help you, you're getting the money" so it's not really an option'. For worker 16 considering a printer's job was pointless: 'I know I would be wasting half-hour of my time, I know I wouldn't get in. He was going to fob me off. I know for a fact that no woman would get in there—they're [pressroom] all the same'.

Further (survey) comments included: 'there is an old boys' club thing in my company—women are excluded. They have also been undermined and moved sideways in the past', or 'I'm female so limited to certain departments. I've got no actual qualifications so can't go into the print side (at 35 I'm too old for an apprenticeship) and I don't play football, rugby or golf!' Craig et al. (1984: 83) noted that even the women's supervisors tended to be male, removing obvious promotion opportunities. There were also occasional examples of machine assistants (male) as well as finishing assistants (female) being denied promotion opportunities.

So while Chaps. 5 and 7 reveal the benefits for craftsmen in work organisation and pay structure, this evidence shows that promotion plays only a limited role in its maintenance. However, as Crompton and Sanderson have argued, women recruited directly into low-level categories 'are likely to stay there … [if these are] sex-typed *jobs*' [original italics] (Crompton and Sanderson 1990: 35). Craftsmen maintained closure (technical and collective power) against semi-skilled men and all women, maintaining a segmented workforce with the co-operation of management in both recruitment and promotion processes. Here it seems, deliberate inactivity, as much as overt resistance, can be a factor in maintaining the status quo (Lukes 2005: 40).

Training Processes

Opportunities for gaining initial qualifications and both maintaining and improving them are of central importance in accessing higher pay in printing. The evidence below reveals the paucity of training provision generally and the role this played in reinforcing occupational segregation, despite the perennial problem of skills' shortages. To put the printing industry record in context, the 2003 National Employers' Skills Survey indicated that '53 per cent of employees had received some training over the previous year … [a level appearing] more-or-less constant since the late 1990s' (Keep 2005: 213). However, Keep acknowledges that distribution of training varied across the workforce, with women part-timers faring especially poorly and that much training was in any case 'informal, uncertified, uncosted and … unrecorded' (ibid.: 214).

Considerable proportions of print workers had no qualifications (35 per cent of men, 66 per cent of women), but the crucial figures are those for the traditional apprenticeship route—City and Guilds qualifications were held by 44 per cent of men and only 5 per cent of women. This revealed the extent of male technical power. Figures for NVQ qualifications at this stage were negligible, providing no route to technical power; the industry was still reliant on older workers with traditional skills. R2(8) revealed that this traditional hold on technical power was under strain and training was of much shorter duration, even for skilled jobs, but there were still no women printers. However, employers were now being forced to introduce a new apprenticeship scheme to address the problem.

Table 6.4 illustrated the short training periods generally available. However, women clearly outstripped men in these categories, and short training periods have a recognised link to low pay (Dex et al. 1994). Where training periods were three years or more, indicating the likelihood of apprenticeship training, men had a clear advantage (25 per cent to only 6 per cent of women). While '[l]ack of "human capital" is not … a generally valid explanation of women's employment position' (Craig et al. 1984: 84) given the variety of ways skill is acquired, it had great currency in the printing industry as an exclusionary strategy, maintaining the efficacy of technical power for male workers.

In this survey, regardless of sex, training was on the job for 80 per cent of respondents. There was no further training for 75 per cent of the sample. This may be why 77 per cent of women and 79 per cent of men thought training was either a very important or important priority for the union.

Employers interviewed acknowledged that there was insufficient training to replace those retiring. Any training was firm-specific, and employer 3's company was planning an online training package that employees could undertake in their own time, potentially undermining technical power, while cutting costs.

Union officials felt there should be no barriers to women's training on the new technology but believed employers would, stereotypically, continue to target men, especially given the uncertain demand for labour during restructuring. Officer 9 thought denying training opportunities was aimed at non-craftworkers and was not gendered.

Table 6.4 Duration of initial training

	Non-response (%)	Less than a week (%)	Less than 1 month (%)	1–12 months (%)	Between 1 and 3 years (%)	Over 3 years (%)	Total
Male	12.9	17.4	18.7	19.0	6.9	25.1	100
Female	18.2	27.4	22.7	20.2	5.2	6.4	100
Total	15.3	22.1	20.6	19.5	6.2	16.4	100

Source: Survey of GPMU members 2002 ($N = 1216$)

Women occasionally got training via ULF[1] projects, but officer 7 thought both employers and union officials needed to give women more encouragement to train. She pointed out that:

> it is often the men who train the women and they don't train them the same as they would a man. They can either train them properly or they can train them to fail and you've got to make those confident that it's not taking their jobs and it's not lowering wages. It's a fear of anybody taking their jobs but it's easy to scapegoat the woman again.

She also explained that it was men who generally had transferable skills so they retained access to new technology, especially as many companies only trained when new equipment came in, just topping up existing skills. Most union officials also said that companies relied on poaching.

Officer 4 made the case for a proper valuation of the existing skills women had acquired, linking collective bargaining and the provision of training that has been noted elsewhere (Craig et al. 1984: 69). Women interviewees generally thought their companies' training provision was poor, especially the older women; often they were expected to pick it up as they went along so never found out the short-cuts, 'it's a muddle along attitude really' (worker 27). Worker 2 explained how she was supposed to do a three-year apprenticeship, but once she had learnt all the processes on the single colour press, they did not continue her training, as agreed, on the four-colour press. She believed this was partly because the printer training her objected to her knowing all the operations on all the machines. This is a good example of 'men's sense of owning the organisation' (Cockburn 1991: 46). Further examples included a woman machine operator turned down for training as a printer in favour of a male cleaner and another who saw her own trainee promoted above her; the latter example is a corrective to officer 9's comment above.

Survey evidence demonstrated how lack of access to training could be an absolute bar to acquiring technical power resources. One woman attested,

> my supervisor thinks I am ready to be upgraded, when he spoke to the manager he said I wasn't ready. I am being made to learn all the machines, which I have already done. I just don't think they want a female grade 1.

Another woman explained, 'When there was a chance of promotion, less experienced men were given jobs to train as main finishers, although I had more experience'. Clearly these examples have nothing to do with ability but reflect systematic undervaluation of women's skills, 'low pay and status of women employees … [being] taken as an indicator of low skill and job content' (Burchell et al. 1994: 181). Men fared better because their apprenticeship training and the lack of fundamental technological change meant they could 'pick up' the knowledge needed to run the new equipment.

One final point captures the effect of attitudes in an industry with a history of apprenticeships (Craig et al. 1984). In interviews with unions and union members, employers were said to be reluctant to train women who may take time out to have and care for children. There is a clear link here with expected working patterns, and the next section takes up this discussion.

Hours of Work and Caring Responsibilities

This section considers the evidence for working patterns and links it to both caring responsibilities and to domestic labour more generally. Beginning with extended working hours, some 79 per cent of workers had overtime opportunities in 2002. Table 6.5 looked at the frequency with which overtime was worked.

Fifty-three per cent worked overtime occasionally and this was regardless of sex, while 30 per cent (34 per cent of men and 27 per cent of women) did overtime frequently and only 15 per cent never worked overtime (11 per cent of men and 19 per cent of women). While unsocial hours working was found to be common among parents (Cousins and Tang 2002), it may also reflect the lower numbers claiming parental responsibility in this survey. This evidence and Table 6.6 contradict research finding that overtime was generally restricted to men in manufacturing industry (Horrell and Rubery 1991).

A further 19 per cent of respondents, not working overtime, answered a question asking if they would be prepared to work it; some 60 per cent said yes (65 per cent of men and 59 per cent of women). This suggested

Table 6.5 Frequency of overtime working, by sex

	Non-response (%)	Frequently (%)	Occasionally (%)	Never (%)	Total (%)
Male	1.1	34.0	53.6	11.3	100
Female	1.2	27.2	52.4	19.2	100
Total	1.1	30.5	52.9	15.5	100

Table 6.6 Extent of extra hours working, by sex

	Non-response (%)	Less than 2 hours (%)	3–5 hours (%)	6–9 hours (%)	Over 10 hours (%)	Total (%)
Male	16.1	15.5	22.9	24.8	20.8	100
Female	22.7	13.1	23.4	26.2	14.6	100
Total	19.4	14.3	23.2	25.4	17.7	100

the vast majority of print workers, regardless of sex, either worked or were prepared to work overtime. Only in the categories containing part-time workers was this not so. In general, overtime was paid at the premium rate (89 per cent of men and 77 per cent of women), although R2(7) and R2(8) suggested this had not always remained the case after the global financial crash. Commonly, part-time workers would not get these premium rates (Horrell and Rubery 1991), and this was confirmed in the survey. Furthermore, only at the longest levels of working hours (over ten hours) were women significantly less visible than men (Table 6.6).

Figure 6.1 shows that 58 per cent of respondents worked shifts (73 per cent of men, 42 per cent of women) and indicated a move towards 24/7 forms of working, later confirmed by R2(7) and R2(8), although one day may be a maintenance day. The union was not always supportive of women's problems with changed working patterns, as worker 18 experienced when annualised hours were introduced, to the detriment of managing her childcare responsibilities.

Working time can be used 'to rationalise or legitimate existing systems of occupational segregation' (Horrell and Rubery 1991: vii) rather than consider alternatives. There was evidence for this in employers' interviews. They were quick to rationalise women's reluctance to work shifts as a reason why women did not apply for skilled jobs: essentially because

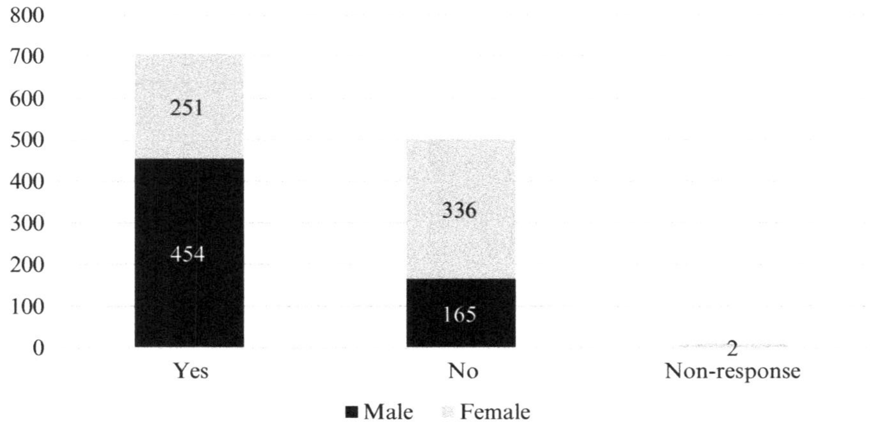

Fig. 6.1 Shiftworking, by sex. Source: Survey of GPMU members (2002) (N = 1216)

of domestic responsibilities. Certainly in the survey women finishing assistants were more likely to work early and late shifts, not the 24/7 cover often required in the pressroom. Union officials also thought the increase in seven-day working was a barrier to women with caring responsibilities.

The majority of workers had taken no childbirth leave (61 per cent of men and 69 per cent of women). In fact less than a quarter of women had taken statutory leave (24 per cent), meaning over three-quarters of women had not. Employers confirmed this, claiming they were rarely asked to provide maternity leave. Women of childbearing age provided numerous examples of the barriers they faced. Legal rights were grudgingly provided but little information about what they were entitled to. Worker 14 took reduced leave because she needed the money but was still penalised on her return, being put on a lower rated machine because her less than three-month absence meant she needed to 'catch up'. While men were generally ignorant of these problems, some had to take holidays because of the inadequate statutory paternity pay. Such barriers may explain why 58 per cent of respondents claimed to have no dependents. In fact only 19 per cent of women had school-age children. There was also a high level of non-response suggesting few had dependents.

Employers interviewed demonstrated a reluctance to adapt to caring responsibilities. Employer 5 remarked that it was difficult to let women come back from maternity leave to work part-time in production—and this was despite the numerous shift patterns he then outlined. He claimed never to have been approached about childcare. Employer 4 relied on informal flexibility, perhaps attempting to ensure maintenance of a stable workforce (Bradley 1986). The multinational utilised continental shifts, which, employer 3 felt, assisted childcare because of the blocks of time not spent at work, failing to recognise the difficulties of accessing childcare for unsocial hours. They offered only statutory rights and did not welcome the flexibility rights that were about to come into force.

Such hostility to caring responsibilities had consequences. Worker 30 said, 'they wouldn't have hire [them] if they'd had childcare difficulties', while worker 29 said, 'they have never ever worked round people with kiddies. You either accept it or you don't'. One example of hostility to childcare affected worker 14 when her son was ill: the company refused to pass on the message for four hours, by which time her son was in hospital. Worker 19 felt lucky because they let her work opposite shifts to her husband to manage childcare, while worker 1 explained how statutory rights could embed occupational segregation. A woman worker in his company trained up to a more skilled job, took maternity leave but was refused part-time work on return so moved to lower skilled despatch work. A minority of men experienced such problems. It appears that women did not try to retain a job in print during early motherhood, apparently unable to exercise any power resources to challenge the status quo. Instead, they left the industry when their children were young and returned when they were older.

The survey provides some context on part-time working: only 12 per cent of women and no men identified this as their working status, and those women tended to be in the lowest pay band. However, one difference between 2002 and the industry in 2017 would be that negligible numbers were casualised then, but R2(6), R2(7) and R2(8) said there was considerable employment of agency and migrant labour in the industry today.

Evidence on the sharing of domestic labour confirmed that women did most of it. But men were more likely to say they shared these responsibilities than women (43 per cent compared to 30 per cent). It was also still primarily women who took time off to attend sick children, but almost half of respondents did not answer this question.

The evidence above suggests that women were still being corralled into traditional roles with employers reluctant to adapt to flexible working that worked for women. However, the numbers claiming no dependents, little call for part-time working and few with school-age children suggested a degree of self-selection by women to manage unhelpful inequality regimes. Employers' economic and positional power, insufficient mobilisation of collective and legal power from the union and inadequate state regulation made equality seem a non-issue—irrelevant and invisible to all but the few who found themselves facing legal claims. 'Proceduralisation' masked unequal outcomes and tended to create an attitude of anticipating discrimination and a certain fatalism that meant it was not worth challenging the way workplace inequality regimes were maintained. The next section considers the union's attempt at greater inclusiveness, an important consideration in challenging workplace practices.

Union Internal Democracy

The Efficacy of Structural Change in the GPMU

Chapters 3 and 4 identified the various problems facing the GPMU: general economic and specific industrial restructuring, alongside the failing power of skill as a bargaining tool, and a hostile political environment, all leading to falling membership. Along with many other unions, the Graphical, Paper and Media Union (GPMU) adopted strategies encompassing versions of both organising and partnership (Daniels and McIlroy 2009), each with a gender dimension. The formation of the GPMU was, itself, a defensive strategy aimed at accessing economies of scale in its administration allied to a desire to end the vicious competition between the Society of Graphical and Allied Trades (SOGAT) and the National Graphical Association (NGA), thereby providing greater scope to increase membership (Gennard and Hayward 2008). The growing inclusion of women in these difficult times was linked to these strategies (Colgan and Ledwith 2002: 1). 'Proceduralisation' through structural change had a paradoxical effect, as will be demonstrated below; while this allowed for limited short-term change, long-term transformational change remained out of reach, leaving the inequality regime largely intact.

The purpose of increasing women's influence in decision-making was ostensibly to ensure that they would gain better wages and conditions in the workplace, and so be more likely to join, and remain in, the union. However for this to happen, men would have to accept changes in workplace hierarchies, prefigured by changes in union hierarchies. This implied a willing surrender of some of men's power in order to increase women's power. Why would any group willingly surrender power? The answer ought to be based on recognition of the underlying purpose of a union—to build solidarity to serve all members' interests. On the contrary, the historical reality of trade unionism rested on sectional interests, especially among craft unions (Hyman 1975). Theoretically, those unions who could not rely on skill around which to build strength would be more open to recruiting broadly, regardless of skill, and focused on pay movements (Turner 1962). But these unions still tended to build their action around discrete male groups, for instance, manual workers in local government often gained bonuses that women workers in local government did not have access to, which later led to significant numbers of equal pay claims (bbc.co.uk, 2013)—controversially, not always with full union support, partly because of the potential damage to all workers in those sectors and, arguably, because central government refused its responsibility to fund the claims (The Guardian 2010). The tendency is to undermine solidarity, especially between white men and groups experiencing disadvantage, such as women.

Traditionally, analysis of union purpose focused on the three methods unions relied on to represent members—mutual insurance, legal enactment and collective bargaining. In other words, understanding of what unions were for was based on what they did (Flanders 1971, originally 1965). Research often concentrated on collective bargaining (ibid.; Clegg 1976; Martin 1992), and this will be discussed in the next chapter, dealing with equality bargaining. Much less was written about mutual insurance and legal enactment, but they are important for understanding how the GPMU responded to employers' tendency to sideline women's equality and for how the union attempted to improve women's representation in the union without upsetting the majority male membership. This approach also explains the failure of structural change to transform the inequality regime.

Analysis of power resources demonstrates the gendered trajectory of trade union purpose. Trade unions are by their nature democratic organisations. While unions' structures and government differ by union, they all encompass varying, but generally substantial, degrees of officer accountability to members (Hyman 1975). In the GPMU all officers at every level were elected; the general secretary was subject to election every five years in accordance with the Trade Union Act 1984. Industrial action and negotiated agreements were subject to ballot. These ballots had been secret workplace ballots until the Conservative legislation in 1984 required them to become postal ballots. In a union such as the GPMU, it is not surprising that the high degree of accountability would lead officers to pay greater attention to the skilled male majority. Employer 1 claimed that skilled workers constituted at least 70 per cent of those covered by the national agreement in general and commercial printing.

As Chap. 5 makes clear, this essentially meant the vast majority of those subject to the agreement were male: very few women, if any, were accorded skilled status in this sector. Similarly, the GPMU, like other unions, often recruited its paid officials from those who were high status and typically male (Hyman 1994). In essence, collective, sexual and technological power resources provided access to *inter alia* positional and personal power resources within union hierarchies. Put simply, skilled men's collective organisation permitted colonisation of positions of control and influence of both the formal and informal 'rules of the game' inside and outside of decision-making bodies.

The power resources available to men in the union encompassed virtually the whole range of those identified in the model. Women's power resources were largely restricted to their sexual and collective power in campaigning for change, with much weaker access to other power resources. Pressures for change coalesced, as Chap. 5 explained, shortly after the formation of the GPMU. The new union, although dominated by NGA officers, had to make some accommodation for the 17 per cent of members who were women (27 per cent of former SOGAT membership). This may not have amounted to much without both the SOGAT women's determination (sexual and collective power) and external pressures from the TUC and Labour Party, both of which realised the necessity for greater inclusivity. Societal change also drove expectations of greater equality for women, especially as they were becoming an increasingly large segment of the workforce (GPMU 1992).

The result was guaranteed proportional representation (GPR), ensuring places for representation for women on all decision-making bodies within the GPMU. This did not just mean committees and conference delegations but also negotiating panels. Such developments were becoming more general throughout the union movement, the impetus coming from the publication of the TUC Charter *Equality for Women Within Trade Unions*. These developments have had extensive analysis (e.g., Pollert 1981; Cockburn 1991; Cunnison and Stageman 1995; Colgan and Ledwith 1996, 2002; Kirton 1999; Kirton and Healy 1999, 2012; Parker 2009; Healy and Kirton 2013). Successive TUC surveys have traced their development throughout the union movement (e.g., TUC 2016). While progress is clear, it remains limited. Although the GPMU is the case study in the current analysis, it is contended that variations on the theme of power resources are likely to hold explanatory potential in most, if not all, trade unions, as long as they are historically specific.

The limitations of the GPMU gender structures are apparent in a number of ways. Women's sexual and collective power resources might be assumed to have delivered GPR, but Colgan and Ledwith (2002) argue that the general secretary was seen as having delivered it although, as argued in Chap. 4, in the circumstances of merger with a union that had much higher women's membership, it would have been hard for the general secretary to resist some form of greater representation for women. However GPR came about, men's power resources were always more extensive and more firmly embedded in the industrial relations of the industry.

An example of the limitations of structural change can be seen in the workings of the executive council (EC). This body was made up of a mixture of lay representatives and full-time, paid branch officers. When GPR was first introduced, it started with the EC elections. As recounted in Chap. 4, the mechanism used was, in some ways, quite progressive because it tried to avoid the tokenism of reserved seats. Instead, women had to stand in the main election, and recourse to reserved seats was only brought into play when insufficient numbers of women were elected in their own right to meet the requirements of proportionality (GPMU 2000). This meant four women needed to be elected to the council. Initially, this resulted in a record number of women standing for election and four women being elected in their own right. However, this had the consequence of denying a seat to a man from the broad left. It appears that men learned the lessons of this result. At each subsequent election, the numbers of women standing fell, and the necessity to rely on reserved seats increased until the requirements of proportionality could not be met at all. Thus, there were no more threats to expectant male candidates in the electoral process, and within six years GPR had been undermined on this key body, thus ensuring an agenda that would not challenge labour market disadvantage (Munro 1999). It is worth noting that the branch with the largest women's membership, Scotland, never nominated a woman for the EC at all, ensuring that their two seats would always be held by men.

Positional and personal power were important here because branch officers, as paid officers, were in charge of circulating calls for nominations for all union positions, including the women's structures. This had two potential outcomes. On the one hand, it meant any branch officer so inclined could ensure that only acceptable candidates got to know about the call for nominations. On the other hand it allowed officers to claim no one was interested if there were no volunteers. As with employers' claims that women just do not come forward for higher-paid jobs, so the union could claim lack of interest for union positions. However, when trying to introduce diversity in long-standing structures, it is generally necessary to encourage and motivate the underrepresented to take part in the change. Lack of action (non-decision-making) meant those branches not disposed to the changes could avoid them. The survey of branches' implementation of GPR (GPMU internal survey results 1998) suggests that there was some element of this in some branches, although others put great effort into making sure the changes worked. The attenuation inherent in decentralised policy implementation is well known (Munro 1999: 129; McBride 2001: 20, 37, 67), but this evidence demonstrates how it can operate. Officer 7 had concerns in this regard, noting that sometimes women were being brought forward just to fulfil the requirements of the rules 'rather than because they're any good … and it doesn't help our sex'. But officer 14 captured the nature of positional and personal power in action:

> some branch secretaries feel that the branch is their dominion and fill the committee with placemen and women. It's patronage, so when nominations come up it's the good boys and girls who get support. They don't circulate the call for nominations to the membership.

The result was that the holding of key positions in the hierarchy with both knowledge, and some degree of control, of the informal elements of union democracy could limit the challenges inherent in GPR and continue to marginalise women's issues. This may be the clearest example of the mobilisation of bias (Lukes 2005). But despite its imperfections, it is clear that GPR was still necessary as officer 4 remarked, 'I would like to think we could get there by being elected by right but I think we are a long, long way from that'.

It is not intended to deny general membership apathy is a barrier to filling union positions. This is a perennial problem for unions. But for women the alienation may be profound. An officer from a more proactive branch said, 'why should they want to travel in one evening a month to sit with ten blokes discussing what's been happening in the branch?' (officer 7). Similarly, officer 3 explained that 'GPR is not any more or less likely to create change. Women have other pressures and a dinosaur outfit [to deal with]. It is equality on … [men's] terms'. However, women's reluctance to engage in union internal politics (Colgan and Ledwith 1996) limited their ability to embed their position once structural change occurred.

Furthermore, the interaction of the women's structures with the general structures could obscure the visibility of women's action on inequalities. The union had encompassed notions of *interim separatism* (Colgan and Ledwith 1996: 164) through the women's committee and conference, but the effects were limited. While other research identified the lack of a direct link to decision-making structures as a problem (Munro 1999: 58), the existence of such a link was part of the problem in the GPMU. Some benefit was evident in raising women's awareness and increasing their activism, but the role of women's structures in policy-making was poorly understood. It was the fact that policy recommendations went straight to the EC for endorsement that hid women's involvement; the credit for introducing new policies went to the EC, not women.

While it may be argued that women were part of the EC, and so their role was not invisible, this raises another issue. Proportional representation would seem the fairest way to ensure representation, but where the disadvantaged group is such a small minority, this mechanism might be insufficient to create more than minor change. For example, officer 13 thought real change was limited because:

> It has been pitched at a level where we were round about what we'd be able to get by traditional methods … I do think we've pitched it at a level where we're were going to get that participation anyway that's why I don't think it's … changed too much.

In comparison, women were previously 44 per cent of the bookbinders' union and remained a substantial proportion of SOGAT. This may be why those unions took more action on issues like equal pay (see Chap. 4). It appears that where women are a critical mass, they can effect more significant change and the tipping point for effectiveness is likely to be substantially higher than 17 per cent. However even in unions with a majority female membership, for example, Unison, women have yet to reach the highest positions and transformative change has still not arrived, despite self-organisation for women and minority groups (McBride 2001; TUC 2016).

Women struggled within this context to be effective. One of the complaints from a number of male officials was that women failed to speak out on the EC. This criticism is laced with irony because, during the period of the research, the EC was generally noted for its lack of debate and tendency to let items go through 'on the nod', especially those dealing with equality. This is exactly the style of meeting women are known to dislike (Briskin 2006). There were also calls from officers for women to really drive the agenda and for a woman national official to have this responsibility. This ignores the reality that women national officials had, in fact, had this responsibility for years but were often stymied by having to operate to a male model of what an officer should be, with long hours of work and no domestic responsibilities (Kirton and Healy 1999; Kirton 1999). Chapter 7 indicates how difficult it could be for women officials where they were lulled into 'doing the job like a man', but apparently still lacked credibility, or felt they were evaluated not just as officials, but specifically according to their sex. It is also notable that only one of the women officials (officer 7) had children and in her experience:

> You have to give up certain things—and hand on my heart I haven't spent, as a single parent, as much time with my children in this job as I might have in another job. I have had to give up quite a lot really and I don't think it's recognised.

In the later research, R2(4), who had a key role in women's equalities, said,

> women were fearful of what I would do. Women don't want to be battle-axes; women don't want to be like me. I would not do it again. I would not want to put myself through that ever. Why should women have to turn themselves into … monsters just to get something done?

There was also the question of attitudes towards women in the industry. Former SOGAT officials tended to think that NGA members had always 'treated [women] with disdain' because they 'never crossed their path' at work (officer 5). While NGA officials rarely mentioned this, officer 13 did acknowledge the lack of dignity and respect accorded women workers, which he identified as due to women's designation as unskilled. Given the contentious nature of what constitutes skill and its traditional absence in the evaluation of women's work (Phillips and Taylor 1980), it appears that women were highly unlikely to ever be accorded respect.

The GPMU: A Reassessment

This scenario seems laden with doom and gloom, but a necessary corrective is to assess what was positive about the changes the GPMU embraced and their potential, even if unachieved. This evidence primarily comes from former GPMU women activists in the later research, allowing them to assess the changes in their old union from a distance and compare to their most recent experiences in Amicus and, subsequently, Unite.

GPR, despite its limitations, gave women a guaranteed opportunity to raise their concerns in a way they had never previously experienced. Many women got places on their branch committees, and this led to greater opportunities at national level on the EC, conference delegations, women's committee and conference and myriad ad hoc meetings, working parties and so on where the union embraced the idea that women should be present. There was also a relaxation of the criterion for access to union training courses. The normal criterion was for members to be local union representatives; for the women, this requirement was dropped in recognition of the need to build women's confidence in becoming active. In this regard, there were courses for women who either had some local role or showed interest in becoming more active. The union relied on branch officers to 'talent spot' for this purpose and encourage women members. There was also a new role for women in the union, liaison officers, which required lay women members to act as the liaison between branch officers, of whom only six were women, and women members facing sexual harassment in the workplace. Training was arranged for these posts.

The women's conference was also important here as it adopted a hybrid approach. It was not a motion-based conference but organised around themes into workshops. The workshops encompassed elements of both training and encouraging women to express an opinion on issues of particular concern to them. The resulting discussions would be formalised into recommendations for the women's committee, debated and accepted at the plenary session at the end of conference. The women's committee would discuss the recommendations and send them in a comprehensive report to the EC for endorsement for action. Women full-time officers, staff and women's committee members ran the training at courses and conferences. Women emerged often much more enlightened, as those interviewees who had attended confirmed. For instance, it emerged at one of the conferences that women had rarely ever seen a copy of the national agreement relevant to their sector. R2(7) summed up her views on these changes, saying the GPMU approach was:

> very positive. There was direction from national level, the women's committee had a network and could tap into regional women. The Conference was important, especially using our own women to teach.

She explained,

> as there were few women at national level it gave other women a chance to have a profile and level of experience. It wasn't a "pull up the ladder" approach but trying to expand … It also had broad political views, the women's network [names two women] were quite strong for women.

Where women had achieved paid official positions, this could be important for women too. Interviewee 7 explained that although they struggled to get women onto the branch committee, they were increasingly getting attendance at the Annual General Meeting. To encourage further activism, this branch also held a pilot scheme of training for a group of women who attended a number of courses to enable them to build knowledge and confidence and also to network with each other. Good results were achieved from this experiment in terms of future activism. This pointed a way forward that was abandoned as the GPMU headed towards merger.

A senior male official, officer 1, also thought that building gender equality was a long-term battle that had only begun in the last few years, 'we need to keep pushing and pushing and keep prodding away'. He felt the issue was not so much one of men's resistance as a normal process where:

> you are always going to get some reaction, some resentment from the more hairy-arsed members but there's also … a lot of goodwill. It's like when you've got a potential dispute with employers and you can't take chapels with you. Nine times out of ten, if you go and meet them and have the argument, you can bring them round. It's much more a leadership question than our members getting in the way of progress. The leadership question is very much reflected at branch level.

This raises the issue of leadership and making change count on the ground. Certainly, there was a feeling of potential not realised, as Chap. 4 demonstrated, there was little policing of the national agreement at local level. But overall, R2(3) indicated that:

> I think, in hindsight what we did was quite good, the structures made a possibility of women moving forward, moving up, being included in all levels of the union, although that wasn't achieved without (laughs) some aggravation.

At national level, responsibility for equalities was the purview of successive women officers. Initially, this was a national official, but they either retired or moved on in the first three years after merger. Then the responsibility moved to the next level down, industrial officer. A break down in relationships then required a novel approach because there were no more women national or industrial officers. This approach broke with union custom in two ways: a staff member was appointed (not *elected*) equality policy adviser, and the remit was extended from women to all equalities. This could not have been achieved without the assistance of a key woman on the EC:

> you have to give Tony Dubbins credit for creating the equality policy adviser. I agreed the job description with Tony Dubbins. It needed to be a mainstream issue, not an academic discussion. Members must directly input. (R2(7))

The increased development and implementation of policy along with campaigning for better legislation and continuing support for legal cases was the result. However, this demonstrates greater attention to mutual assurance and legal enactment than collective bargaining. Chapter 7 discusses the detail of the latter, but it should be noted that much of the bargaining resulted in either statements of intent on the employers' side or explicit restatements of the law within the agreement. Pay was not a key element. In fact, in the mid-1990s the women's committee was forbidden to discuss the grading structure in the national agreement (Colgan and Ledwith 2002: 172). Furthermore, Chap. 7 explains how even landmark legal decisions could be ignored, a clear example of mobilisation of bias (Lukes 2005) delivered through positional and personal power, with symbolic power supplementing them (Dawson 2014). In essence, sectional interest of the dominant men's group trumped greater solidarity and the inclusivity expressed in the organising strategy.

But there was some disappointment with the subsequent mergers that has put in perspective what the GPMU achieved in a relatively short time. Of the Amicus merger, there were some disquieting experiences recounted. For instance,

> the Amicus conference that I attended people were saying that, just ordinary delegates were kind of making outrageous sexist comments, racist comments, homophobic comments. It really was a kind of culture shock … and the shocking thing was that most people didn't react. There was us frothing at the mouth, I mean including male colleagues, not just the women. So that was, I suppose, quite an eye opener. Apart from that things weren't, you know, although lots of people had obviously fought quite hard within GPMU to get positive structures, people fought for it. But then afterwards we did seem to take a step back. (R2(3))

Similarly,

> Although can have a good moan about the GPMU from an industrial and political view they were well in advance. Amicus was a bit of an icy shock. Some of the attitudes over low pay, class, equality etc.—it felt like we were faced with the same arguments as the 1970s. [names senior official] who was 'Left' of a sort, was sexist—it was an unfortunate interregnum … but control of the 'Left' in AEU to break the EETPU/AEU might was a success and UNITE/T&G completed that break. (R2(4))

For women, generally, mergers were a problem: 'Every merger the profile of women has got smaller … in the Amicus merger everything was disbanded. In Amicus there was no visibility, no agenda' (R2(7)). However, it was pointed out that within the new GPM/IT sector of Amicus, the senior officer 'tried to keep the flag flying' for women's and equalities' representation (R2(7)). Sectoral GPR was maintained as far as possible (R2(6)).

The Unite merger was viewed with more optimism but some frustration that things are still not back to where they were under the GPMU. While the sector as a whole had less influence, for equalities:

> It has a broader class position and so is less susceptible, but not immune, to sectoral … differences … there is potential for progress

But:

> there's committees at every level, lots and lots of structures [in Unite] … but it's trying to get people involved and not smothered by bureaucracy or controlled. (R2(4))

R2(3) had the most experience of trying to use the structures and found a number of problems with Unite, although she recognised the goodwill in what was being attempted:

> It's [equalities] a kind of afterthought … not necessarily because people are anti-women but they just don't see it as being part of what they need to do.

Despite the union being lay member led, there were a number of problems with this approach, including the lack of activities taking place at weekends rather than weekdays, given the problems with time off work for activists. Furthermore,

> the only way you get information is from, there is an equality report given at the regional committee, but if you're not on the regional committee you're never going to see that. So obviously, because the union is so large, there's whole swathes of people who are never going to get that information. And there doesn't seem to be any kind of processes to cascade that down. There's certainly nothing on the Unite website … well only in the sense of equality legislation. (R2(3))

Paid officials felt left out. Where the GPMU had used local officer knowledge to spot potential activists and actively seek to involve them in the union, this is not acceptable in Unite because it is an expressly member-led union. The effect is to weaken the structures that do exist and actually weaken accountability for equalities work. Some of this is because of Unite's hybrid structure—partly regional based and partly sectoral based. 'There is a dislocation, which means the bulk goes to the regional committee. NISC [sector] decisions don't get down to lower levels.' (R2(7)).

Finally, R2(5) felt that, despite all the structures, she was not 'sure any of them [the three unions—GPMU, Amicus, Unite] made a lot of difference to workplace equalities. Essentially, power structures stayed the same despite structural change, and greater inclusiveness remained remote from the shopfloor (Cunnison and Stageman 1995).

Conclusions

The account of the GPMU's venture into structural change demonstrates both the potential and the limitations of widening participation in union government. In appearance the GPMU was a union predicated on considerable member involvement. As argued in Chap. 4, however, it actually suffered greater member apathy, moving from exclusive aristocracy towards 'popular bossdom' (Turner 1962). Equality structures may have partially bucked this trend, as demonstrated above, but the small proportion of women in membership, and the persistence of a craft mentality, limited its impact.

In terms of union strategy, greater recourse to organising clearly increased the proportion of women coming into the union. GPR was probably the most important element of structural change for giving these members influence. But it is an example of the paradox of 'proceduralisation'. While GPR clearly improved the visibility of women and their issues, it also contained them through its limited ability to disrupt embedded power resources, which used 'proceduralisation' to the advantage of the powerful. To the extent that it assisted women's positional power, it began the process of increasing their personal power but not sufficiently to challenge men's dominance of the 'rules of the game'. For women to have achieved this would have required them to abandon their disdain for organisational politics (Colgan and Ledwith 1996) in order to pursue a long, transformative agenda (Cockburn 1989) and to form alliances across groups, including men.

While women gained access to decision-making bodies, their sexual and collective power resources were insufficient to do more than dent the positional and personal power of men whose access to control and influence continued to rest on their technological, sexual and collective power resources. The misdirection of 'proceduralisation' made it hard for women to identify precisely where the blockages were and easy for the men to blame the women for failing to make use of structural change. The limitations of internal change provided limited scope to affect the bargaining agenda and tended to push women's interests in the direction of mutual insurance, in the support of legal cases, and towards legal enactment, in campaigning for better legislation. This is important because, as will be seen in Chap. 7, it helped to cement the view of equalities as a social, rather than industrial, issue.

Crucially, the failure to embed women's equality into union structures at a more effective level can be argued to have undermined solidarity at a time when sectional interests were becoming less useful in protecting members. Skill (technological power) was waning, but the union was reluctant to acknowledge this.

These practices and processes demonstrate the maintenance of inequality regimes in both the workplace and the union. The effects on both women and the union as a whole are discussed in the next chapter.

Note

1. Union Learning Fund (ULF) provides government funding for specific training projects.

References

Acker, J. (2006). Inequality Regimes: Gender, Class, and Race in Organizations. *Gender and Society, 20*(4), 441–464.

BBC. (2013). Birmingham City Council to Settle 11,000 Equal Pay Claims. Retrieved November 30, 2017, from www.bbc.co.uk/news/uk-england-birmingham-24383352.

Bingham, C. (2016). *Employment Relations: Fairness & Trust in the Workplace*. London: Sage Publications Ltd.

Bradley, H. (1986). Technological Change, Management Strategies, and the Development of Gender-Based Job Segregation in the Labour Process. In D. Knights & H. Willmott (Eds.), *Gender and the Labour Process*. Aldershot: Gower.

Briskin, L. (2006). Victimisation and Agency: The Social Construction of Union Women's Leadership. *Industrial Relations Journal, 37*(4), 359–378.

Bruegel, I., & Perrons, D. (1995). 'Where Do the Costs of Unequal Treatment for Women Fall? An Analysis of the Incidence of the Costs of Unequal Pay and Sex Discrimination in the UK. *Gender, Work and Organisation, 2*(3), 113–124.

Burchell, B., & Rubery, J. (1994). Divided Women: Labour Market Segmentation and Gender Segregation. In M. E. Scott (Ed.), *Gender Segregation and Social Change*. Oxford: Oxford University Press.

Burchell, B., Elliott, J., Rubery, J., & Wilkinson, F. (1994). Management and Employee Perceptions of Skill. In R. Penn, M. Rose, & J. Rubery (Eds.), *Skill and Occupational Change*. Oxford: Oxford University Press.

Clegg, H. A. (1976). *The System of Industrial Relations in Great Britain*. Oxford: Basil Blackwell.

Cockburn, C. (1989). Equal Opportunities: The Short and the Long Agenda. *Industrial Relations Journal, 20*(3), 213–225.

Cockburn, C. (1991). *In the Way of Women*. Hampshire and London: Macmillan Press.
Colgan, F., & Ledwith, S. (1996). Sisters Organising – Women and Their Trade Unions. In S. Ledwith & F. Colgan (Eds.), *Women in Organisations*. Basingstoke: Macmillan.
Colgan, F., & Ledwith, S. (2002). Gender, Diversity and Mobilisation in UK Trade Unions. In F. Colgan & S. Ledwith (Eds.), *Gender, Diversity and Trade Unions: International Perspectives*. London: Routledge.
Collinson, D. L. (1988). *Barriers to Fair Selection*. London: HMSO.
Collinson, M., & Collinson, D. (1996). It's Only Dick: The Sexual Harassment of Women Managers in Insurance Sales. *Work, Employment and Society, 10*(1), 29–56.
Connell, R. W. (1983). *Which Way Is Up?* London: Allen and Unwin.
Cousins, C., & Tang, N. (2002). *Working Time Flexibility and Family Life: An Analysis of the 'Households, Work and Flexibility' Surveys*. Conference Paper, Gender Research Forum.
Craig, C., Garnsey, E., & Rubery, J. (1984). *Payment Structures and Smaller Firms: Women's Employment in Segmented Labour Markets*. Research Paper No. 48. London: Department of Employment.
Crompton, R., & Sanderson, K. (1990). *Gendered Jobs and Social Change*. London: Unwin Hyman.
Cunnison, S., & Stageman, J. (1995). *Feminizing the Unions*. Aldershot: Avebury.
Daniels, G., & McIlroy, J. (2009). *Trade Unions in a Neoliberal World: British Trade Unions Under New Labour*. London: Routledge.
Dawson, T. (2014). Collective Bargaining and the Pay Gap in the Printing Industry. *Gender Work and Organisation, 21*(5), 381–394.
Dex, S., Lissenburgh, S., & Taylor, M. (1994). *Women and Low Pay: Identifying the Issues*. Research Discussion Series No. 9. Manchester: Equal Opportunities Commission.
Dickens, L. (2005). Walking the Talk? Equality and Diversity in Employment. In S. Bach (Ed.), *Managing Human Resources*. Oxford: Blackwell Publishing.
Dickens, L., & Hall, M. (1995). The State: Labour Law and Industrial Relations. In P. Edwards (Ed.), *Industrial Relations Theory and Practice in Britain*. Oxford: Blackwell.
Flanders, A. (1971, originally 1965). Industrial Relations: What Is Wrong with the System? In A. Flanders (Ed.), *Management and Unions*. London: Faber and Faber.
Gennard, J., & Hayward, G. (2008). *GPMU: A History of the Graphical, Paper and Media Union*. London: Unite the Union.

Goldthorpe, J. H. (1974). Industrial Relations in Great Britain: A Critique of Reformism. *Politics and Society, 4*, 415–452.

Graphical, Paper and Media Union (GPMU). (1992). Women's Representation on the Executive Council, Report to the Executive Council.

Graphical, Paper and Media Union (GPMU). (1998). Internal Survey Results.

Graphical, Paper and Media Union (GPMU). (2000). Rules of the Graphical Paper and Media Union, 50–51.

Grieco, M., & Whipp, R. (1986). Women and the Workplace: Gender and Control in the Labour Process. In D. Knights & H. Willmott (Eds.), *Gender and the Labour Process*. Aldershot: Gower.

Harris, C., & Morris, L. (1994). Households, Labour Markets and the Position of Women. In R. Crompton & M. Mann (Eds.), *Gender and Stratification*. Cambridge: Polity Press.

Healy, G., & Kirton, G. (2013). The Early Mobilization of Women Trade Union Leaders – A Comparative Perspective. *British Journal of Industrial Relations, 51*(4), 709–732.

Healy, G., Rainnie, A., & Telford, J. (2002). The Resilience of Gender Segregation in the UK General Print Sector. Unpublished Conference Paper.

Horrell, S., & Rubery, J. (1991). *Employers' Working-Time Policies and Women's Employment*. London: HMSO.

Howell, C. (1995). Trade Unions and the State: A Critique of British Industrial Relations. *Politics and Society, 23*(2), 149–183.

Hyman, R. (1975). *Industrial Relations: A Marxist Introduction*. London: Macmillan Press.

Hyman, R. (1994). Changing Trade Union Identities and Strategies. In R. Hyman & A. Ferner (Eds.), *New Frontiers in European Industrial Relations*. Oxford: Blackwell.

Keep, E. (2005). Skills, Training and the Quest for the Holy Grail of Influence and Status. In S. Bach (Ed.), *Managing Human Resources*. Oxford: Blackwell.

Kirton, G. (1999). Sustaining and Developing Women's Trade Union Activism: A Gendered Project? *Gender, Work and Organization, 6*(4), 213–223.

Kirton, G., & Greene, A. (2016). *The Dynamics of Managing Diversity* (4th ed.). London: Routledge.

Kirton, G., & Healy, G. (1999). Transforming Union Women: The Role of Women Union Officials in Union Renewal. *Industrial Relations Journal, 30*(1), 31–45.

Kirton, G., & Healy, G. (2012). 'Lift As You Rise': Union Women's Leadership Talk. *Human Relations, 65*(8), 979–999.

Liff, S. (1995). The Industrial Relations of a Diverse Workforce. In *Industrial Relations Theory and Practice in Britain*. Oxford: Blackwell.

Lukes, S. (2005). *Power: A Radical View*. London: Palgrave Macmillan.

Martin, R. (1992). *Bargaining Power*. Oxford: Oxford University Press.

McBride, A. (2001). *Gender Democracy in Trade Unions*. Aldershot: Ashgate.

Munro, A. (1999). *Women, Work & Trade Unions*. London: Mansell.

Noon, M. (2007). The Fatal Flaws of Diversity and the Business Case for Ethnic Minorities. *Work, Employment and Society, 21*(4), 773–784.

Parker, J. (2009). Women's Collectivism in Context: Women's Groups in UK Trade Unions. *Industrial Relations Journal, 40*(1), 78–97.

Phillips, A., & Taylor, B. (1980). Sex and Skill: Notes Towards a Feminist Economics. *Feminist Review, 6*, 79–88.

Pollert, A. (1981). *Girls, Wives, Factory Lives*. London: Macmillan.

Pollert, A. (1983). Women, Gender Relations and Wage Labour. In E. Gamarnikow et al. (Eds.), *Gender, Class and Work*. London: Heinemann Education Publications.

Printing and Publishing Industry Training Board (PPITB). (1982). *Equalising Job Opportunities for Women in Printing and Publishing: An Attitude Survey Report*. London: Printing and Publishing Industry Training Board.

Purcell, K. (1979). Militancy and Acquiescence Amongst Women Workers. In S. Burman (Ed.), *Fit Work for Women*. London: Croom Helm.

Selznick, P. (1969). *Law, Society and Industrial Justice*. New York: Russell Sage Foundation.

South-East Region Trades Union Congress (SERTUC). (2012). *Swimming Against the Tide*. London: South-East Region Trades Union Congress.

Stanko, E. A. (1988). Keeping Women In and Out of Line: Sexual Harassment and Occupational Segregation. In S. Walby (Ed.), *Gender Segregation at Work*. Milton Keynes: Open University Press.

The Guardian. (2010). The Most Hated Lawyer in Britain. Retrieved November 30, 2017, from www.theguardian.com/society/2010/may/01/stefan-cross-female-pay-birmingham.

Trades Union Congress. (2016). *Equality Audit*. London: Trades Union Congress.

Turner, H. (1962). *Trade Union Growth, Structure and Policy*. London: Allen and Unwin.

van Wanrooy, B., Bewley, H., Bryson, A., Forth, J., Freeth, S., Stokes, L., & Wood, S. (2011). Workplace Employment Relations Study: First Findings. Retrieved January 18, 2018, from https://www.gov.uk/government/uploads/system/uploads/attachment_data/file/336651/bis-14-1008-WERS-first-findings-report-fourth-edition-july-2014.pdf.

7

Wage Leadership: The Continuation of Unequal Pay

This study has focused on factors affecting the stubborn refusal of pay differentials to budge significantly—that is, until recently when the general trend in pay has been for the pay gap to close because men's pay is falling, rather than women's pay rising (Rubery and Rafferty 2013), although even this has stalled (Fawcett Society 2017). This can be seen as a falling away of men's traditional power resources that does not assist women so much as it concentrates power even further in the hands of employers and shareholders.

To understand this discussion of pay requires reference to the labour process elaborated in Chap. 5. Jobs and pay were inextricably linked within the national agreement that governed pay in general printing until 2010. The pay structure was and generally remains simple, much as job segregation is. The terminology used is Class 1, Class 2 and Class 3. These equate to skilled, semi-skilled and unskilled basic pay rates. It does not take account of the cumulative effect of extra payments of various kinds. Many of these have their origins in the nineteenth century (see Chap. 3). The importance of this agreement cannot be overstated as it acted as a benchmark, even for companies that were not members of the British Printing Industries Federation (BPIF, originally BFMP) employers' fed-

T. Dawson, *Gender, Class and Power*, https://doi.org/10.1057/978-1-137-58594-3_7

eration (Roe 2003). The effect of grading structures is often to institutionalise pay relativities (Wootton 1955; Colling and Dickens 2001), and this was the case in print.

The mechanisms of this agreement are important in understanding how actual differentials came to be so much wider than the apparent narrow differentials in the agreement. Both unions and employers used this narrowness of basic pay differentials at different times to prevent further narrowing and even to argue for increasing them. The agreement worked as follows: Class 1 denoted 100 per cent of the basic rate, Class 2 was 92½ per cent, and Class 3 was 87½ per cent. This does, indeed, seem narrow. In practice extra payments made these differentials little more than fiction. A Graphical, Paper and Media Union (GPMU) senior employee explained how the pay packet of a Class 1 worker was made up. The first element is the machine extra payment. This calculation could produce an extra payment varying between £20–100 per week for printers and £13–38 per week for skilled bindery workers. The printer's extras were split if there was also a second printer on the machine, for example, three-fifths to two-fifths. Machine assistants (Class 2) also got extras based on what the printer's extra was. Extras were added to the grade (basic) rate listed in the national agreement.

There could also be a payment for a running changeover to ensure continuous operation during the shift changeover. This could accrue an additional ¼–½ hour's pay. House extras would be added, and they covered anything additional the chapel had been able to negotiate with the employer. Shift pay was then paid on the sum of these extra payments. From this an hourly rate was calculated on which overtime was paid. Women were always unlikely to gain much from this, given they worked fewer extra hours, did not run machines that accrued extra payments and were often confined to single-gender chapels because of occupational segregation. According to R2(8), following the global financial crash (GFC), these extras were often consolidated or bought out, although just recently, strong chapels were beginning to claw back some elements of the pay package.

One final aspect of the agreement is important to understand for the later discussion in this chapter. In 1988 it was agreed that there would be a minimum earnings guarantee (MEG) paid at the Class 2 basic rate. This meant that anyone working a full 37½ hour per week would get that basic rate. This was obviously useful for women as they were, in the main, paid at Class 3 basic rate. This obtained until 1994 and the change made at that time is discussed below.

The survey evidence from the original 2002 research demonstrated the results of this in practice. Table 7.1 is based on figures from two separate questions in the survey. The first asked respondents to identify their class position in the agreement or whether they operated under a different system. If they chose the latter option, they were asked to give an indication of the level of their pay grade. The *N* = number is higher than total respondents because there was a small number who ticked both questions (73). It is clear that women were concentrated in the pay grades for the unskilled occupations with men overwhelmingly concentrated in the pay grades for skilled workers, whether or not they followed the agreement. The 24 per cent of women in Class 2 probably reflected the success of SOGAT in negotiating away Class 3 in some of the larger companies during the 1980s (Roe 2003), although this apparently made little difference in effect (Craig et al. 1984). Certainly the women in Class 2 were most often found in companies with over 100 employees and would still not be entitled to most of the extra payments.

An understanding of the structure of pay in the industry lays the groundwork for understanding the discussions of pay determination that follow. The rest of the chapter first considers wage determination post Second World War and then looks at the situation after the formation of the GPMU, before examining pay processes in detail.

Table 7.1 National agreement grades by sex and notional indication for those not following the agreement

Pay grades	Women (%)	Men (%)
Class 1	15.9	60.1
Class 2	23.6	8.9
Class 3	16.1	3.2
Near top	2.9	6.2
Middle	11.5	10.6
Near bottom	13.9	5.9
Non-response	16.1	5.1
Total	100	100

Source: Survey of GPMU members 2002 (*N* = 1289)

Pay Determination Between 1950 and 1991

After 1945 it was not uncommon for printing firms to have retained a marriage bar, and the bookbinders had to issue a circular to branches instructing them not to remove women's union cards when they got married, even if they left the trade (NUPB&PW 1953). R2(2) recalled that she had become a 'jobber' after she got married, meaning she had to do casual work, which she continued for the rest of her time as a print worker. The recommendation to treat married women equally, also contained in the circular, was often ignored (Nicholson 1994). Clearly women's pay could be affected by such restrictions.

Conversely, the level of women's membership at this stage brought a greater focus on their pay and closing the pay gap within their union. As a result, in the 1950s there was a clash of strategy between the craft and non-craft unions over differentials. The bookbinders wanted to achieve a women's rate of 75 per cent of the craft rate, but the craft unions wanted to widen differentials. At this time all women members 'irrespective of their actual skill levels—were grouped under the 'women's rate' at the bottom of the grading structures' (Gennard and Bain 1995: 487). This came to a head in the 1955–1956 negotiations where four of the craft unions insisted that 'any subsequent increase granted to non-craft employees could result in an application for a further increase in craft rates' (SOGAT 1955). Neither of the non-craft unions would accept this position and stood firm, even though talks came close to breaking point on several occasions (Gennard and Bain 1995). This meant that leapfrogging claims were not just a problem for employers but for women too. However, by 1956 women, using their collective and sexual power resources, had reinforced the claim for a 75 per cent rate, thereby establishing women's pay as a legitimate part of the bargaining agenda.

In London, women's rates had moved from 46 per cent of the craft rate in 1939 to 59 per cent after the 1959 dispute, while in the provinces the respective change was from 44 per cent to 60 per cent (ibid.). In the process, the policy on gender differentials was watered down to 66 per cent at the urging of the craft unions (Child 1967; Gennard and Bain 1995). Subsequently, there was an attempt to close differentials in 1960 when the bookbinders' union agreed to push for women to get the same increase as Class 4 men. However, at one large employer, HMSO, rates were still only around 55–58 per cent of the lowest male rate, 84 per cent of women earning less than £18 per week, while no craftsmen earned so little, and only 4 per cent of male SOGAT members were in this bracket (Jones 1969).

The 1959 dispute also saw the bookbinders raise the issue of extra payments (usually applied only to 'men's machines) for one group of women. The employers would only concede this on condition it established no new principle of paying extras to women (Gennard and Bain 1995). The resulting agreement also turned the tide against craft control by establishing principles of efficiency and productivity (Roe 2003) that would not bear fruit until the 1980s and 1990s, when they would be used to demand that all wage settlements be self-financing.

There was a gradual acceptance of the need for equal pay legislation after the non-craft merger, leading the Society of Graphical and Allied Trades (SOGAT) to campaign for equal pay from 1966 onwards (Gennard and Bain 1995). SOGAT recognised the problems with the Equal Pay Act 1970 in that it did not take account of equal value and would thereby leave job segregation intact. As employers were starting to introduce women into typesetting, there was the added danger that a pool of cheap labour would be utilised, but the NGA did not heed this warning (see below).

Equal pay demanded union negotiators' attention when the 1972 SOGAT conference challenged the differential between Class 4 (the lowest men's rate) and the women's rate. The NEC responded that they had made some limited progress in reaching equal pay agreements with employers but that the 'greatest problem was male prejudice on the part of SOGAT members' (ibid.: 491). The union's aim was that women's work should be properly valued and 'slotted into the male grading structure' at the appropriate points and that there should be 'complete freedom of access to jobs' (ibid.:491). In 1974 SOGAT put in a claim as part of the national negotiations with the BFMP and NS[1] for the removal of the women's grade and that women should move up into Class 3. The employers rejected this claim, along with the additional claims for maternity leave, wages and holidays. Rejection of the claim led to a strike. The outcome was that all the union's objectives were obtained (SOGAT 1974a, b). Not all unions accepted this approach, and some were more concerned with maintaining differentials (Gennard and Bain 1995). For this reason the dispute had been confined to SOGAT alone.

Women moved up in stages to Class 3 wages[2] by October 1974. This, of course, coincided with the introduction of new laws on sex discrimination and equal pay. But it also supported SOGAT's aim to increase the wages of the low paid to stop poor wages dragging down the higher paid, an argument that had to be won with the membership (SOGAT 1974c). Similar arguments raged about women's right to have access to all the jobs in the industry too (SOGAT 1975a, b).

The NGA finally obtained equal pay for women operating small offset machines[3] in August 1975 (Gennard 1990). They were not in the least dismayed by the introduction of the Sex Discrimination Act and the Equal Pay Act that came into force in 1975 because they believed that employers would now be more reluctant to hire women than men as they would be more expensive. However, not all union members were happy with adopting a policy that 'we would have deaf, dumb, blind or crippled men rather than accept women' (NGA 1976). The union's London Region challenged this policy in both 1974 and 1976 eventually succeeding in changing it to one of accepting people for the job on the basis of their skills and receiving the rate for the job (Gennard 1990). As Gennard emphasises '[o]ne result of these policies is that equal pay is not a problem for the NGA as it is for some other trade unions' (Gennard 1990: 446). This disingenuous statement ignores the barriers women have faced in getting access to jobs classified as skilled. In fact printing was one of the industries where little was done between the passing of the Act in 1970 and its implementation in 1975 (TUC 1973). The Act's provisions required that women not earn less than the minimum male rate under collective agreements, and this bare minimum was all that was conceded under the BPIF agreement.

In the 1980s SOGAT continued to push for improvements to the wage structure that would particularly affect women workers, including abolition of Class 3, with some success.[4] The BPIF rejected this claim, 'arguing that Class 3 still had relevance in the industry and was best dealt with at local level' (Gennard and Bain 1995: 475). This probably reflected the fact that, although SOGAT had had some success in removing Class 3 from agreements with the bigger companies, smaller companies still largely retained the three grades and dominated the BPIF's membership. The NGA would not support the removal of Class 3 but 'made a proposal which, whilst going towards meeting the Sogat (82) position, ensured that NGA interests would be taken into account in any discussions on changes in relativities and differentials' (Gennard 1990: 418). SOGAT also tried to get pro rata wages and conditions for part-time workers, but the BPIF rejected this claim for a number of years until it was finally conceded to the GPMU in 1994[5] (Gennard and Bain 1995).

The influence of extra payments had continued from the nineteenth century. In 1980 a Class 3 worker made the point that new technology usually made the machine operator's (Class 1) job easier but not so the Class 3 members 'who feed, stack, and parcel the end-product [and] are physically working harder each year' (SOGAT 1980). But it was the Class 1 workers that received extras for the machines they operated.

Chapter 4 identified the extent of women's flexibility, which the national agreement ignored. The claim made at J & C Moore in Liverpool was groundbreaking as it proved it was possible to equate Class 3 jobs with Class 1 jobs. In this case the comparators were table hands (women, Class 3) and guillotine operators (men, Class 1). This claim took years, and the employers' defence was that it was the bargaining structure that had created the anomalies (Gennard and Bain 1995). The tribunal rejected this. Subsequently there was an out-of-court settlement which included substantial back pay and a job evaluation scheme to ensure there was no discrimination in the pay structure.

Although this was a landmark case (SOGAT 1990), Ledwith was premature in suggesting that employers and unions were 'reassessing their positions on the class and pay structure' (Ledwith 1991: 54). It had little effect on the industry generally, revealing the weakness of the law in removing endemic sex bias (Dickens 2005) and the resistance of unions and employers to challenging gendered bargaining practices. Other successful claims followed (Gennard and Bain 1995). But some employers, for instance, the *Financial Times*, imposed pay cuts on the male workers rather than raise the women's rates (ibid.: 493). However, the effect of the law was only to help a few individuals. Instead employers and union officials were generally protectionist, undertaking limited action to avoid litigation against their respective organisations. Limited success was apparent in women getting into jobs that were traditionally male preserves including in national newspapers and general print, but these were rare occurrences and, in reality, little changed.

Legal power was achieved by women's campaigning (their sexual and collective power resources), but it is clear from the above that it was relatively easy to avoid its consequences. Women's personal and positional power resources were too weak, given their relative isolation, to ensure that they would make use of their legal rights. Meanwhile, men's and employers' personal and positional power resources ensured that they could avoid making women aware of these rights and rely on the weakness of individual rights to allow landmark decisions to whither on the vine. It was not so much collusion as a coincidence of interests, protecting gender differentials and status, while employers benefited from women's cheap labour. The change required to make the agreement concordant with the law did not require anything more than a different interpretation (symbolic power) of the job tasks within categories, no change to the agreement was necessary. As it stood, women could do any number of tasks classified as Class 3, and their cumulative effect was ignored; the J & C Moore case challenged this. In the event it was a missed opportunity to build cross-gender solidarity.

This case spanned the period when the GPMU was formed and its consequences were felt within the new union. The next section considers in more detail the actual mechanisms of negotiation and how these continued to marginalise the gender pay gap (GPG).

Pay Determination Post-1991

The first GPMU women's conference was told that the root cause of women's problems was the failure of employers and workers to accept that women should have equal access to all jobs in the industry (GPMU 1991). The general secretary stated that 'Only 10% of employees in origination are female, less than 2% on machines and 12% in finishing', focusing on the problems of occupational segregation, but he went on to explain that, apart from access to jobs, equal pay should be tackled industrially, forcing employers to pay the rate for the job (ibid.). But women were already getting the rate for the job—a rate that incorporated sex bias and itself needed challenging to address the systematic undervaluation of women's work. As neither of these issues was seriously challenged, this left women 'caught between a rock and a hard place'—undervalued for existing work and with little prospect of accessing better-paid jobs.

A year later the women's committee reported to the next conference that the union was opposed to job evaluation schemes as a hindrance to collective bargaining. Furthermore, the union would only co-operate with tribunal equal value cases 'providing all other industrial avenues have been exhausted' (GPMU 1992a: 6). But industrial solutions were rarely pursued, and a limited number of individual legal cases were the only result, reinforcing the idea that gender issues were not industrial issues and ensuring that they remained marginalised and relatively invisible. These arguments were subtly guiding women away from challenging the GPG (an exercise in personal and positional power). Job evaluation threatened to diminish bargaining power, further enhancing employers' positional, personal and economic power resources. But given the extent of job segregation, it is hard to see how jobs could be compared without some form of evaluation procedure. Class and gender potentially conflicted here.

The lack of progress on the GPG saw it raised again and again at women's conferences.[6] These entrenched problems formed the backdrop to national negotiations throughout the GPMU's existence. The rest of this section looks in more detail at, first, national bargaining, then local bargaining and the translation of the national agreement into actual wages.

National Bargaining

The first big industrial challenge for the merged union was the 1993 dispute with the BPIF over the national agreement (Roe 2003), and this had special significance for women. The employers framed their offensive in terms of attacks on 'social issues' (a pernicious interpretation of equality), that is anything not conforming to the traditional wages, hours and holidays agenda, and a desire to take advantage of what they saw as union weakness in merging during the recession. The GPMU accordingly did not include 'social issues' in the claim that year (Roe 2003), reinforcing the invisibility of women's concerns.

The Federation's stated aim was 'to see the differential of Class 3 workers fall from 87.5% to 70% [of the Class 1 grade rate]' (GPMU 1993: 78) and also cuts to overtime and shift premia. The offer that emerged after negotiations was regarded as risible, and a campaign was mounted on a company-by-company basis to achieve £6.50 on Class 1 and a day's extra holiday (ibid.). By the summer over half the membership was covered by the union agreement (ibid.) and over 65 per cent by September (GPMU 1995: 8). A national agreement was re-established in 1994 (ibid.).

Although the employers' desire to widen differentials was defeated, the union needed to get the agreement re-established nationally, and there was a price to pay. Any widening of differentials, albeit small, could not but widen the GPG. In re-establishing the agreement, the union conceded a temporary widening of differentials, meaning the MEG, previously set at the Class 2 rate, was to be set at a new rate between Classes 2 and 3, virtually creating a new grade—20 years after the abolition of grade 4 and the women's rate with the introduction of the Equal Pay Act, as officer 4 forcefully recalled. Furthermore, on an ongoing basis the MEG would only be upgraded by the rise for Class 3 (ibid.), ensuring gradual and continuous widening of differentials. In 1996 the union carried out a survey on a number of outstanding questions, including differentials, and there was little support for widening them, although it 'exposed a fault-line between former NGA and SOGAT sections of the union' (Roe 2003: 255).

Differentials confirm status (Phelps Brown 1977), so the apparent narrowing of basic pay differentials was threatening to skilled men (Cockburn 1983), providing opportunities for employers to make cost savings—and the Federation had planned this assault. The BPIF Strategy Review 1992 began the offensive against ostensibly overpaid (relative to other industries) women print workers (Roe 2003). Employer 1 provided a candid description of their approach. It is worth quoting his explanation in full:

> Differentials are a function of the labour market. Historically the supply of skilled workers was controlled, latterly through the Joint Apprenticeship Panel ... joint regulation. Competition for skilled workers led to a rise in

> the rates and no competition for unskilled workers, so [skilled workers pay] rose at a faster rate. [In the 1990s] we tracked back over 10 years and found differentials widening steadily, probably as a result of skill shortages. There's only so much money to spend, the wage pot. There were problems recruiting skilled workers, not unskilled, and rates were higher than the local labour market [he refers to regional factors, especially in the north]. We wanted to spend money on the skilled rather than the unskilled. For the union, this was a problem. The steps we took to uncouple the MEG and Class 2 were cautious. We hoped that over time it would widen out.

Union weakness and probably the 'customary expectations of relative or absolute living standards' (Rubery 1997: 341) dominated decision-making, despite all the structural change aimed at giving women an influence. This development supported the employers' aim of full cost recovery (FCR) in all agreements, it was something they could 'claw back'.

Officers did not seem to consider that the employers' claim that women were paid too much in print might not stand up to scrutiny. But when directly questioned, officer 13 admitted,

> I think they're [the employers] just comparing other industries where there's unskilled workers on lower pay. I don't think it's a … measured comparison or even one that they try to justify on the basis of similar or same work being undertaken, it's just vis à vis skilled and unskilled.

The pragmatic decision to concede widening differentials supported the majority male group of members. It also demonstrated the shared assumption, between union officials and employers, that differentials were too narrow.

Negotiating Processes at National Level

A look at the mechanics of negotiations at national level is instructive. The development of shared understandings between key negotiators appears to marginalise bargaining issues of importance to women.

National negotiations were an annual event, and Walton and McKersie's (1965) seminal analysis of bargaining behaviour, although it did not consider the issue of gender, is useful in uncovering the subtleties of gendered behaviour and decision-making within these negotiations. Their focus was on negotiation as a social process where 'two or more social units are attempting to define or redefine the terms of their interdependence' (p. 3). They developed a model comprising four sub-processes and a synthesis that demonstrated the complexity of negotiations. These sub-processes comprised distributive bargaining (a zero-sum game, where if one side gains, the other loses), integrative bargaining (a problem-solving approach that has potential for gain on both sides), attitudinal structuring (the development of the desired relationship patterns between negotiators) and intra-organisational bargaining (negotiators' tactics for balancing the conflicting interests within their constituencies without compromising the ability to reach agreement with their opponents). Their analysis was based on manufacturing industry and sought to identify the range and effect of tactics used in the interaction of these sub-processes.

Negotiating behaviour can result in smoother negotiations and more realistic expectations where the parties know each other and understand the pressures on the opposition (ibid.). Shared experiences in the industry are likely to assist such an understanding and may make tacit signs easier to read (ibid.). Looking at the composition of the respective national negotiating panels and the characteristics of the key negotiators emphasises the importance of these relationships.

Employers have their own agenda to pursue, which is affected by their experiences and attitudes. Employer 1 was a professional federation employee of over 20 years standing and with a good understanding of the federation's members and was their former (recent) chief negotiator. Generally, the employers' panel was all male and only occasionally included a female panel member. There was sometimes a woman on the support staff for the negotiations.

There were few women at any point within the national officer cadre of the GPMU; for a time there were none. Those that did get to that level tended not to come from print manufacturing but from white-collar occupations within the union or industry. Conversely the men came from time-served backgrounds and Class 1 jobs, albeit they all tended to have over 20 years' service, so their direct experience of shopfloor life was somewhat distant. Women officers were conscious of the difficulties of trying to do their job and not neglect equality. For instance, officer 3 was concerned that since becoming a national officer she was 'doing the job like a man … not hunting for it [inequality] specifically. If I come across it I will deal with it. I will say, "what about the women" as low paid', but the workload and time pressure meant 'there is no time to go cruising', so unless it was raised with her, she was unable to do much.

It was from this group and from the executive council (EC) that the union's panel members were drawn—four from the EC, of which one or more was usually a full-time branch official[7] and, latterly, a woman lay member was also included. From the national officials, the general secretary (GS, chief negotiator) and deputy general secretary (DGS) were always present plus additional national officers with responsibility for the agreement would also be called on for specific agenda items. The author, as equality policy adviser (EPA), became part of this panel specifically for equality issues. The agenda consisted of a list of items that would be split into different negotiating sessions. The GS would deal with non-equality issues and the DGS, who had overall responsibility for equalities, with equality issues. Various combinations of panel members supported these senior officers, and the whole panel might assemble for particularly important matters such as pay.

The chief negotiators of each panel had been negotiating together annually since the 1980s, as well as meeting outside of this forum many times. Attitudinal structuring established the union's chief negotiator as tough and highly respected, a heroic leader (Briskin 2011). What the make-up of the panels suggests is that many on the employers' panel, and most on the union's panel, had developed a shared understanding of the industry and of their expectations of negotiations, based on their familiarity with each other and the industry. This served 'the crystallizing of the bargaining agenda' (ibid.: 314) and the attempt to gain a shared outlook among panel members that can encourage a straightforward approach to negotiations, with little dissent from the direction provided by the chief negotiators. Hegemonic groups often dominate collective bargaining (Dickens 1997; Ledwith 1991), and with little involvement of anyone likely to have a different outlook, this hegemony is evident here.

The effect of such homosociability on negotiating processes was considerable, not so much questioning the intent to include equality, as to raise the spectre of groupthink dominating outcomes. In this context, women struggled to acquire credibility. As officer 1 made clear, the lack of a production background 'has created some difficulties with other officers. There's been overt pressure on them to run the equality agenda … [rather] than taking up industrial issues and political issues'. This highlights an apparent distinction between equality and industrial issues that suggests the former is of lesser importance, mirroring employers' approach to 'social issues', potentially an indicator of attitudinal structuring. There was, therefore, a powerful form of delegitimisation on two fronts: questioning the ability of women to handle negotiations, especially of industrial issues (see Willis 1977) and questioning whether equality issues should be on the agenda at all (Acker 1998). The employers compounded this approach with one final tactic, interpreting the recent increase in equality claims, rather cynically, as solely driven by the need for survival, rather than principle, 'So employers are not surprised that most minority groups are on [the unions'] list. The union is a business that needs to grow' (employer 1).

These tactics provided the scope for the coincidence of interests over equalities to result in marginalisation of issues important to women. Employers could look to reducing costs, while the union could preserve a semblance of status and higher pay for male workers. Access to key power resources, collective, technical and sexual for men, economic for employers, allowed both groups to use their positional and personal power resources to dominate negotiating discourse, the discourse itself being an indicator of symbolic power.

Power becomes most apparent in the negotiating arena. Various tactics identified in Walton and McKersie's analysis support a gendered reading, such as letting items fall off the agenda (Walton and McKersie 1965: 334–335), which can serve to avert the negotiation of equal pay, similar to Lukes' (2005) third dimension of power. Similarly, informal contact that may build shared understandings between negotiators (Walton and McKersie 1965: 63, 112) may lead to an interpretation of what is appropriate for the bargaining agenda and its prioritisation, a form of attitudinal structuring. Furthermore, intra-organisational bargaining raises questions about how internal conflict over gender pay differentials might be handled. Sub-processes can interact too (Findlay et al. 2009). In this regard, equal pay is a mixed issue with both distributive and integrative potential. Clearly increasing pay for women is distributive, but both unions and management share a desire to avoid litigation, thereby opening up possibilities for integrative bargaining.

Before negotiations commenced both sides had to take soundings from their memberships to identify bargaining issues. Men, especially time-served printers, who were less likely to identify equality issues as a priority, dominated this intra-organisational process. The next stage was the informal meeting between the chief negotiators plus one other from their teams. For the union this was the GS and the DGS; for the employers it was the full-time BPIF negotiator and his deputy, or perhaps a senior manager, often from a large company. Such meetings are common in formal negotiations and act to map out each party's expectations (Walton and McKersie 1965: 63, 312). While it is unlikely that equality issues were denigrated in these discussions, their placement early on the agenda suggested that they could be disposed of quickly, leaving 'new money', unsurprisingly, as the final item and top priority. Such processes have been identified as inimical to challenging existing pay structures (Colling and Dickens 1989).

One tactic that can serve employers interests is to appear to make concessions, which actually cost nothing (Walton and McKersie 1965: 334–335). For instance, accepting a new clause in the agreement that simply restates existing legal rights. Similarly, clauses to tackle injustice, such as equal opportunities clauses may be no more than a statement of intent (Ledwith 1991). Only their utilisation can make them valuable concessions.

The tactic of designating equality issues as 'social issues' (Acker 1998) could be used to 'stonewall' on improvements to legal entitlements within the agreement, thereby convincing members these issues are just too difficult to achieve (Walton and McKersie 1965: 325). It is likely to be particularly effective where one party is in a weakened position, as the GPMU was following the reforms to trade union legislation (Smith and Morton 2006; Daniels and McIlroy 2009). The print agreement was based on negotiating for Class 1 increases that would automatically trigger a rise based on the set differentials for Classes 2 and 3. This is called wage leadership (Barrett 1980; Cockburn 1983); the ostensible intention is to drag up the lower paid by using skilled workers to set the rate. Cockburn noted that there was a tension in this approach: union members might also resent 'the lower ranks for "climbing on the skilled man's back"' (p. 128). Officer 14, nevertheless, felt this was a useful tactic that had benefitted the lower paid. But officer 1 held a more nuanced view,

> there's been substantial changes. I think the basic sort of traditional differentials have narrowed very considerably [but] on the ground I'm not so sure … because … the women that are left are still in the main, in the finishing area and the main areas of higher pay are certainly in the press room these days and there are few women in the press room area.

This comment represents a generally held view among male officials—access to jobs, not valuation of women's existing skills was the issue for them. Officer 9 exemplifies this approach. He said 'My own view is not so much that there is such a huge need for improving women's rates of pay but a need to move [them] into better jobs'. But women had little chance of getting those jobs, so there was limited potential to challenge the status quo.

However, many issues of importance to women did make it on to the agenda. Guidance for handling harassment and bullying was included in the agreement from 1992 onwards, as was the clause requiring equal intakes of male and female trainees (GPMU 1992b). The latter was as ineffective as the move away from apprenticeships to modular training (Ledwith 1991), seen during the transition as reflecting hostility towards non-craftworkers, especially women (PPITB[8] 1982: 14).[9] The number of harassment and bullying grievances remained high throughout the 1990s, despite the guidelines and recourse to the law was often necessary to resolve disputes (GPMU 2002). The equal opportunities clauses had been in the agreement since the 1980s, but there is little evidence that they were ever used. Equal pay was never on the GPMU's agenda, although SOGAT had a better reputation for challenging low pay (see Chap. 4), and, for a time, there was an edict that the women's committee could not discuss the agreement, although this was later rescinded (Colgan and Ledwith 2002).

Women's influence remained limited at the pre-negotiation stage and was compounded in negotiations. Symbolic power, available to union officials and employers, focused attention on the traditional class struggle, preventing gender solidarity as employer advantage increased and wage leadership failed along with men's technical power. While women's issues were more visible than ever, in practice they quickly became invisible as they were shunted off the bargaining table, through processes that treated them as marginal to the greater struggle.

Local Bargaining

The agreement's role in benchmarking pay rates in the industry made it particularly important at local level. It set minimum rates traditionally, and procedures for the rules of engagement, with local chapels free to engage in second-tier bargaining. While claims would start within the chapel (workplace union branch), handled by chapel or maybe branch officials, increasingly national officials would be called on to participate. The latter's background has already been discussed above. However, the make-up of branch officials was no different. Out of 100 branch officials, only six were women. At chapel level, women's committee surveys indicated that only about 11 per cent of representatives were women. Both figures are way below requirements for GPR, but these were two areas not covered by those rules and would have been difficult to enforce given how hard it was to find people willing to be lay officials anyway.

Employers interviewed for this study provided a small cross-section of management. Three were male and one was female. The male managers were all members of the GPMU and apprenticeship-trained, which was common throughout the industry (Craig et al. 1984) and remains so for line managers (R2(8)), although one was a commercial apprenticeship. Employer 3 had no printing industry experience and was a professional HR manager. All the men had between 22 and 40 years' experience in the industry, more in production than HR, although in small companies these roles may be combined. Employer 3 had been in post for ten months but had previously worked in manufacturing industry and was the only employer interviewee representing a multinational company.

The other companies were small or medium-sized. The employers' federation generally represents the small and medium-sized companies, but in this instance the multinational was also a federation member.

It was clear that the male managers' experience mirrored that of those at national level. Conversely, employer 3's progression was less assured. Her previous employers had not employed any women at board level, and shopfloor employment had reflected traditional, gendered occupational segregation. While some male colleagues were paternalistic, others resented her promotion, suggesting it was the result of positive discrimination. She saw workplace attitudes as restricting women to, what are seen as, 'appropriate jobs' rather than total exclusion and progress would continue to be slow.

Of those union officials interviewed about local bargaining, six were national officers and four local. Of these, four were women, two national (one retired) and two local. The male officials' background largely mirrored that of the FTOs, and in common with them, only two had worked with women. As with many of the men, their progression seemed part of a natural process, often arising from preferment by an existing male official.

Conversely, none of the women FTOs had done an apprenticeship. Officer 4 had come from the shopfloor (bindery), while officer 7 had come from an office job and had previously worked for the union and been encouraged to stand for office. Both of these women were below 50. Officer 4 had no choice but to go into the bindery: 'you didn't get the opportunity to become a printer … you had to have an NGA card and [they] vetted potential members. When I started there were no women printers at all'. When she eventually became an FTO she said:

> you would actually have to prove yourself in that position and I felt for a long, long time that I was being judged firstly [by] the fact that I was a female, secondly then, how well I did the job as an officer and that was more so … with the … former NGA craft area.

It was the retired woman national officer who had encouraged her to take a union position. So, all the women officials had similar experiences, having to prove themselves against the grain of natural progression. In fact, while many of the men identified women officer's aptitudes in traditional ways (Briskin 2006), as caring, tolerant, better able to identify discrimination and so on, officer 1 made the point that because women officers have come mainly from a white-collar background, they are not always able to identify with the production areas … 'this has created some difficulty with the other officers'. Here lies the source of women officers' lack of credibility and an example of symbolic power. Officer 1 explained that it would be easier for them in the public sector because 'in our industry you don't get [women] with experience of the machine room issues … [unlike in the public sector where] … they have more experience across the board'.

Women accepted some difference in approach. Officer 7 explained that 'women officers tend to think of the impact on home life of a course of action more than a male officer—for instance, where there was a change in shift'. She also felt that:

> men and women identify the same issues but prioritise them differently … blokes will be going in straight for money but you'd probably look at the fairness of it and how it's going to be worked in the long term … you'd just work it through differently.

Officer 4 said that:

> they don't look for it [discrimination]—they assume everything's right and you've got to look beyond the norm … be more lateral thinking with it. Don't assume that companies do what they say they are doing. Male officers tend to accept what they are told. They want to get in and out—not all of them but … they don't look, they accept it and you can't do that.

This is consistent with the general themes of women's leadership styles (Briskin 2006) and of men's domination of the informal agenda at local level (Cunnison and Stageman 1995; Kirton and Healy 1999).

The attitudes apparent in this evidence suggest a continuation of the coincidence of interest, whereby low pay for women is seen as natural and collusion is unnecessary (Phelps Brown 1977). This 'naturalness' is the essence of Lukes' (2005) third dimension of power.

Bargaining Processes at Local Level

It was at local level that employers were most likely to seek FCR. Conversely, any chapel that felt it had the bargaining power would seek to squeeze whatever concessions it could out of employers.

For employers the business case for equality was important. This was especially so in relation to parental rights and benefits. Legal rights had recently been extended, and employer 4 said, 'I wouldn't stop employing women if they were right for the job [but] … there would be a cost to the business … it depends on the impact … how manageable it is'. While employer 3 thought that employers lacked awareness of the business case, employer 1 said,

> equal opportunities is not an evangelistic cause but makes good commercial sense. It is relatively easy to do … The customer base is changing … buyers may be female, the product may be aimed at women, so the sales force especially needs to reflect it and ultimately the production side too.

But collective bargaining may fail because:

> where there is a decline in membership … it is challenging for managers because Britain is based on an adversarial system—you both argue your damnedest and a compromise is reached. But without it, it is up to management to decide how much to pay etc., and without the other side they could get it wrong and create distrust or an adverse reaction … good trade unions can influence the market in this way. [But] managers, even women, have to juggle pressures. If the level of turnover is low, as it is in many print companies, there is less pressure to address the equal opportunities agenda however important the managing director may think it. The business case in itself is not sufficient because of other pressures.

This raises two interesting points. The first suggests there was still a need for trade union representation as individual employers were often not 'equipped to design new policy "solutions" (Grimshaw et al. 2001: 30). Additionally, while the arguments for the integrative potential of equality bargaining were recognised, the reality may reflect "the pursuit of organizational self-interest … [and therefore] only a partial, selective approach"' (Dickens 2005) that is undermined by the pressures of production. The business case is, therefore, fundamentally flawed (Noon 2007).

Could legal solutions assist? Union officials saw the role of the law in delivering equality as a poor substitute for bargaining but despite its weakness, it was a necessary bargaining tool. But the employers were largely untroubled by its existence because they had never been subject to litigation. In fact, employer 3, from the multinational, exemplified their insouciance. Decentralised bargaining with little monitoring is known to exacerbate inequalities (Colling and Dickens 1989), but still this was their approach. At that time, campaigning for pay audits was high, but 'jobs are well scoped. I don't feel we would be vulnerable … we're not intending to do a pay review'. This is despite the job segregation that could throw up potential equal value claims. Instead their approach was to try and keep the union at arms' length on equality. Even the maternity laws resulted in few requests and so were manageable. Employer 4 demonstrates a view more commonplace among traditional male managers, 'finishing, in your mind, is of lesser value so you don't pay as much'.

But if the law is an inadequate tool, then equality bargaining assumes greater importance. A key factor here is union weakness. By 2002, when this research was conducted, the union was only two years away from merger with Amicus and its membership losses were beginning to bite. Survey respondents, of whom only 9 per cent of men and 7 per cent of women were chapel representatives, attest to their focus on protecting existing wages and conditions from cuts. The result was that many interviewees thought raising grievances was pointless, 'you know they're [management] not going to do anything. They pass the buck and you go round in a circle' until things die down (worker 11). Where traditional chapel control was evident, the situation was better, 'it's the old school. The FoC's [shop steward] been around a long time. I'm very impressed with the whole thing' (worker 3). Worker 1 illustrated the greater difficulties of even getting a basic rise, while anything extra tended to be discretionary and not consolidated (worker 2).

There was a lack of knowledge about the agreement among members, generally only noticed when the ballot papers came round. Members also rarely saw the agreement, according to officer 7, and she felt that there needed to be more advice on how to avoid institutionalised differentials. Women interviewees tended to feel management offered them no more than a 'take it, or leave it' option, rather than bargaining, so they felt obliged to accept what was on offer. Members often seemed to feel disenfranchised, too. As worker 4 explained, 'it is very rare to have a union meeting … we never know about talks'. This may account for lower voting turnout. The feeling that union power was slipping away also led to a feeling that 'companies are getting the upper hand … chapels are less frequently asked to be involved in negotiations … decisions appear to be made more and more at the top'. Despite 52 per cent of men and 46 per cent of women claiming they were consulted, this may only refer to the ballot. Only 38 per cent of men and 29 per cent of women claimed to follow the agreement, anyway, with 28 per cent women and 26 per cent of men claiming the employer decides on pay rises.

Men were generally much more aware of how things got done on their behalf, while older women were least aware. However, where the union had only recently organised companies, knowledge was greater. For instance, worker 30 described the active democratic structure at her company where the union had only been organised for two years.

In this context, the contraction of local bargaining limited the scope for pursuing equal pay, 'there's not as much house bargaining done now' and it requires much more education of chapels on equality to raise issues (officer 12). This concurs with research demonstrating that local equality bargaining is rare, tends to be reactive and follows a narrow agenda (Colling and Dickens 1989; Heery and Kelly 1989). These characteristics were all apparent in printing. None of the employers interviewed had been approached on equality bargaining, and, in fact, there was little evidence of the agreement being policed locally. Where local bargaining did occur, it tended to focus on machine extras, not something generally applicable to women.

Again, the explanation proffered for lack of gender equality was occupational segregation and its effect on bargaining,

> it's nothing to do with men and women because the industry breaks down into departments [i.e. whether the pressroom feels solidarity with the bindery]. I don't detect a reluctance to support women because they are women.

> There are traditional splits. Where men and women are together in a department I don't see any lack of will to support each other. It is tribal … old unions or demarcation lines like print or origination. (officer 9)

However, the suggestion that these divisions were accidental misunderstood the gendered nature of the traditional splits common in workplaces (Cockburn 1994; Harris and Morris 1994; Rubery and Fagan 1995). It neglected an understanding that women 'are poorly paid for what they do' (Crompton 1997: 44). Officer 4 felt strongly about this because she had experience of challenging the classification of jobs in the national agreement. But, as with the J & C Moore case, above, this may challenge the sacrosanct nature of the rate for the job (Musson 1954) and could be seen as an attempt to use symbolic power by, and on behalf of, women, although not fully realised. Officer 3 captured the reason behind women's lack of success in challenging traditional bargaining, '[men] think of workers and trade union members as men. If you are not part of something in the first place there is a perceived separateness'. Coming late to the industry made women invisible in bargaining processes and weakened their access to power resources. Even when they did try to create change, this weakness made it relatively easy to incorporate their demands in a way that was harmless to the status quo. This was also reflected at local level where even the more progressive branch officers tended to see equality as a social, rather than industrial, issue. Officer 11 said, 'Virtually every claim we do we put forward social provisions, that's mostly to do with the equality of women'.

None of this presumes there is no attempt to make gains for women or that there is necessarily bad faith. Officer 13 discussed two successful equal pay claims, the only ones of which he had experience:

> The thing that's always … been encouraging is that when we have had a real issue, then the men have been supportive, maybe not totally supportive in the way the women would have liked. But even when there was clearly some personality clashes … the men were prepared to take a lower increase to help the women come up. So I think, faced with the situation, men do suddenly recognise that they have to do something to change and they have to make a contribution.

It may also be relevant that this official was seen as stereotypically NGA, and this might convince other craftsmen of the need to make such sacrifices in the face of potential legal challenges—he may have been believed where others would be ignored. These cases were also examples of where threats of legal action provided individuals or small groups of women with power resources to challenge the status quo. However, complying with the law was as much about protecting the union from prosecution for misrepresentation as it was about providing justice for women workers, so officer 13 had a strong motivation to carry the men with him where legal action was a serious threat.

Clearly to achieve gender equality would require strong union solidarity. Member interviewees reported that there was more a 'look out for yourself' attitude' than previously. At best it was a 'look after your section' attitude,

> they had a big meeting … and talked about all sticking together and uniting. Back on the shopfloor we ended up with two separate agreements—the press hall avoided continental shifts but the bindery got them. So now it will happen … They've always kicked us when we were down. They [printers] think they're a step above us. (worker 21)

Conversely, worker 28 was optimistic about the possibility of solidarity. However, she qualified this statement in relation to the printers. She said, 'printers seem to be a breed on their own. Printers think of themselves as different—everybody's friendly—everywhere you go printers are a breed apart'.

Pay determination mechanisms reveal the centrality of occupational segregation as a means of ensuring differentials. They were used during the formation of trade unions to ensure technical ability was paramount in bargaining. Subsequently, they were used to rationalise the lack of change, despite the weakening of wage leadership and technical power. Whatever arguments women put forward, they seem to be doomed to remain caught between a lack of access to better-paid jobs and a systematic undervaluation of their work. However, the evidence advanced in Chap. 5 meant that women were losing their jobs to technology and the strong basis of their claims for equal value, based on the number of Class

3 tasks undertaken, was rapidly overtaken by events. The next section discusses the outcomes of these pay processes and of the eventual failure of the national agreement in 2010.

Pay Outcomes

Evidence of women's low pay informed high-profile investigations into women's pay generally (see, e.g., Davies et al. 2000), and the matter continues to gather headlines (see, e.g., Fawcett Society 2017). Davies et al.'s (2000) novel approach considered the lifetime penalty for women workers. Even women with no children could expect to earn £241,000 less than a comparable man during her working life, raised to £381,000 if she had two children.

The trends in women's pay in print can be traced through official statistics. Official statistics reveal that in 1970 manual women earned 47.8 per cent of men's weekly earnings in printing and publishing and 53.6 per cent of their hourly earnings. The Equal Pay Act seems subsequently to have raised women's weekly earnings to a high of 65.6 per cent of men's earnings in 1976 (70.8 per cent for hourly earnings). Hourly earnings did not reach their high point until 1977 when women earned 72.8 per cent of men's hourly earnings (64.5 per cent of weekly earnings). Thereafter women's earnings fall back through the 1980s, rising a little in the 1990s to reach 64.3 per cent of men's weekly earnings by 2001 (67.5 per cent of hourly earnings). Thereafter there is a discontinuity in the figures that prevents comparison.[10]

In printing the last time it was possible to calculate manual earnings, it revealed that full-time women earned 68.7 per cent of men's full-time hourly median earnings (31.3 per cent GPG) and 64 per cent of their weekly median earnings (36 per cent GPG) (New Earnings Survey 2002). The most recent figures suggest a big fall in the pay gap, as the 2016 revised figures give a GPG of 9.77 per cent on weekly gross median earnings (5.76 per cent gross median hourly earnings). Figures disaggregated by occupation tell a different story. Comparing like with like, the GPG for printers was 36 per cent and for bindery workers was 14.3 per cent gross weekly median earnings. (31.5 per cent and 8.71 per cent, respectively, for gross median hourly earnings). However, taking account of occupational segregation requires comparing printers with bindery workers, by gender. These figures reveal a GPG of 30.1 per cent gross weekly median earnings and 27.39 per cent gross median hourly earnings (ONS 2017a, b, c).

Survey and interviewee evidence indicated that all non-craftworkers were suffering as the craftworkers pulled away from them, the employers concentrating resources on the 70 per cent or more who were craftworkers. However, some women felt constrained by the relatively recent national minimum wage (NMW) legislation. As a woman survey respondent explained, employers manipulated pay to ensure they never got above the NMW, 'we have to do piece rate every day to have £14 taken off our money to make our wage £4.10 per hour. What happens in October if the minimum wage goes up—more work off us to have more money taken off us'.

Union weakness solidified in the curtailing of benefits evident in the survey, including cuts to machine extras, call out money, paid meal breaks, house money, health insurance, sick pay and, occasionally, pensions and the first signs of individualised pay. These are the results of FCR and work intensification, 'we get awarded the pay award from the BPIF. I think they're members [the company, but the employers always want to] … claw back the cost … but we're always improving really. They're always adding things onto the machine and getting extra money for this equipment—this principle has got to stop' (worker 3). Unsurprisingly, survey evidence indicated considerable dissatisfaction with pay (59 per cent of women, 51 per cent of men), with an almost identical proportion feeling their pay was unfair. But almost two-thirds of respondents indicated little opportunity to improve pay.

So what do the processes revealed in this chapter mean for actual pay? Tables 7.2 and 7.3 are only intended to be indicative of the level of gender pay differences.[11] While 21.7 per cent of women respondents earned £200 per week or less basic pay, only 1.6 per cent of men were in this position. Conversely, only 1.3 per cent of women earned over £500 per week basic pay compared to 15.5 per cent of men. Women predominated in the £210–£350 per week basic pay bands (62 per cent), which accounted for 40 per cent of men, a significant proportion of

Table 7.2 Basic pay

	Non-response (%)	<£50 (%)	£51–100 (%)	£101–150 (%)	£151–200 (%)	£201–250 (%)	£251–300 (%)	£301–350 (%)	£351–400 (%)	£401–450 (%)	>£500 (%)	Total (%)
Male	6.4	0.2	0.0	0.6	0.8	9.0	14.7	16.7	16.6	19.5	15.5	100
Female	8.1	0.0	2.7	6.2	12.8	31.6	17.6	12.6	4.2	2.9	1.3	100
Total	7.1	0.1	1.3	3.3	6.7	20.4	15.9	14.6	10.3	12.0	8.4	100

Source: Survey of GPMU members 2002 ($N = 1216$)

Table 7.3 Gross pay

	Non-response (%)	<£50 (%)	£51–100 (%)	£101–150 (%)	£151–200 (%)	£201–250 (%)	£251–300 (%)	£301–350 (%)	£351–400 (%)	£401–450 (%)	£451–500 (%)	>£500 (%)	Total (%)
Male	8.4	0.0	0.0	0.5	0.3	5.8	8.1	13.8	13.5	18.5	2.9	28.2	100
Female	16.3	0.2	2.0	6.1	7.4	21.3	16.1	13.3	8.6	4.7	0.5	3.5	100
Total	12.3	0.1	1.0	3.1	3.8	13.2	12.0	13.6	10.9	12.1	1.8	16.1	100

Source: Survey of GPMU members 2002 ($N = 1216$)

whom were at the higher end. Moving on to gross pay indicated even greater differences. The £200 per week or less category fell to 15.7 per cent for women but 0.8 per cent for men, while the over £500 per week category rose to 28.2 per cent for men and 3.5 per cent for women. The middle section of pay dispersal, i.e., £201–£350 per week, contained almost 51 per cent of women but only 27.7 per cent of men.

Looking at Table 7.4 may provide some explanation for the gender pay gap revealed by Tables 7.2 and 7.3. A significantly higher proportion of women than men did not answer this question suggesting either that fewer women received extras or that they did not understand the make-up of their pay. Apart from this, only in the categories for bonus pay and performance-related pay did women outnumber men, but, even then, this was marginal.

Well over double the proportion of men compared to women received machine extras (29.8 per cent and 12.4 per cent, respectively), while almost double the proportion of men received shift pay compared to women (61.2 per cent and 32.9 per cent, respectively). Over three times as many men proportionately as women received house extras (12.9 per cent and 4.2 per cent, respectively), and more than double received call money, for call outs at short notice (20.3 per cent and 9.6 per cent, respectively). There was also a difference of over 20 per cent in overtime payments (men 74.2 per cent and women 53.6 per cent), surprisingly, considering the number of women claiming to work overtime (see Chap. 6).

Given men's greater earning potential, it is unsurprising that in Fig. 7.1, 55 per cent of men identified as the main breadwinner compared to only 15 per cent of women. A further 29 per cent of men were sole earners compared to 26 per cent of women. Conversely, women were overwhelmingly identified as neither main breadwinner nor sole earner (53 per cent) compared to only 14 per cent of men and may, therefore, only be earning component wages (Siltanen 1994). In common with some writers (Rubery 1997; Acker 1998), interviewees generally thought the main breadwinner was a thing of the past, but these results, and the comments of a few interviewees, suggested there were barriers to its elimination. For instance, worker 7 explained, 'we work

Table 7.4 Extra payments

	Non-response (%)	Machine extras (%)	Bonus pay (%)	Merit money (%)	Shift pay (%)	Overtime pay (%)	House extras (%)	Sick pay (%)	Holiday pay (%)	Call money (%)	Performance-related pay (%)	Total (%)
Male	4.0	29.8	7.4	5.5	61.2	74.2	12.9	58.3	76.8	20.3	1.4	100
Female	10.9	12.4	7.9	2.0	32.9	53.6	4.2	51.4	69.4	9.6	2.9	100
Total	7.5	21.5	7.8	3.8	47.4	63.8	8.8	54.9	73.2	14.9	2.2	100

Source: Survey of GPMU members 2002 (*N* = 1216)

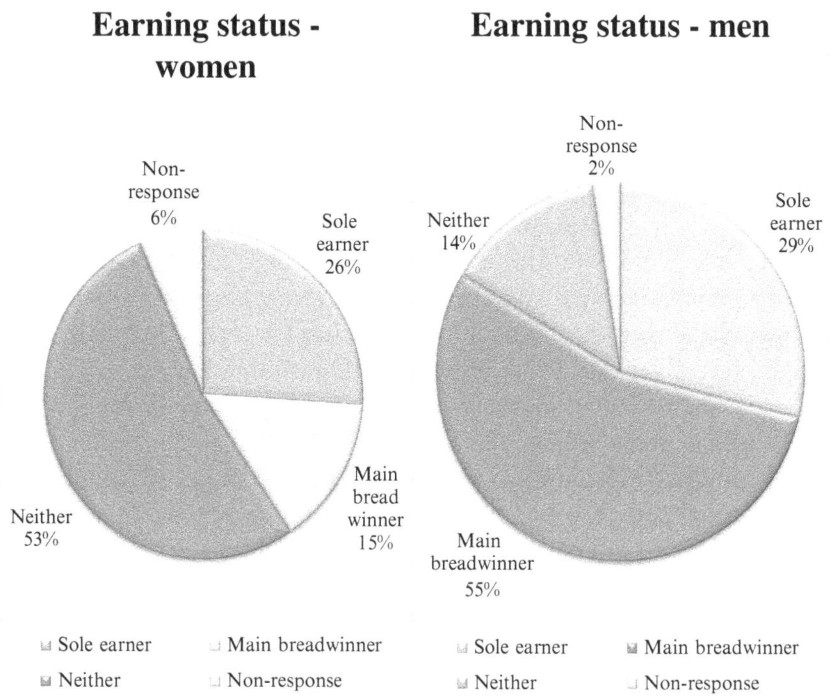

Fig. 7.1 Comparison of family/earning status, by sex. Source: Survey of GPMU members (2002) (*N* = 1216)

three shifts, so it would be very, very difficult for my wife to go and get a job to be the main earner while I'm doing three shifts. I'd love it if she could, so I could get off three shifts'.

Conclusions

Historically, SOGAT had higher women's membership and thus gave more attention to women's pay and conditions post Second World War. They consistently challenged the GPG. While there was some resistance from within their ranks, the main blockage came from employers and male craftworkers. Employers wanted to contain costs, and male craftworkers were concerned with erosion of pay differentials that would undermine their status (Phelps Brown 1977). The employers' economic power and the craftworkers' technical, sexual and collective power was decisive but also gave both parties access to crucial positional and personal power resources when it came to pay processes.

Craftworkers tried to avoid reckoning with the issue of equal pay by claiming they had no problem with it: there was a rate for the job. This attempted to wilfully misdirect enquiry away from the effects of occupational segregation on pay. Women rarely got into craftworker jobs. When women challenged this through campaigning for legal protection, they got the Equal Pay Act 1970 and the Sex Discrimination Act 1975, but these led to an adaptation of pay structures, where women simply occupied a new de facto 'women's rate'.

Pay processes, ostensibly, addressed equality issues, although not directly equal pay. However, the outcomes were generally no more than a statement of intent (Ledwith 1991) or restatements of legal rights. Given that few women ever saw the national agreement (see Chap. 4), it is not surprising that they did not challenge this. Within negotiations, symbolic power was used to interpret equality issues as 'social' (Acker 1998) and women officials as not quite up to the mark (Willis 1977), delegitimising both, rendering them relatively invisible and irrelevant.

All this happened in a context of weakened trade unionism, exacerbated by coercive state power. Unions could not afford to offend the majority male membership by challenging their pay and status—and, in any case, many officers shared the views of the industry with male trade union members and employers. In these circumstances, it is hardly surprising that the pay gap has remained so resilient. The next chapter draws together issues of work organisation and pay processes to present a more complete picture of the barriers women face in the workplace.

Notes

1. For a many years the provincial newspaper employers' federation used to negotiate jointly with the British Printing Industries Federation (BPIF), formerly the British Federation of Master Printers (BFMP).

2. The women's grade and Class 4 were removed at this time creating the agreement that lasted until 2010 with Classes 1, 2 and 3.
3. See Glossary.
4. Roe (2003: 152–153) lists a number of large companies that conceded this claim following the BPIF's humiliating defeat in the 1980 national wages dispute.
5. From the outset, the terms of the 1994 agreement covering part-time and temporary workers was probably unlawful as it denied part-timers the MEG under the agreement. This stipulated that if you did the standard 37½ hours work, you would not earn less than Class 2 pay. Any deviation from this and you could be paid Class 3 if that was your normal rate. As the majority of workers affected by this clause were women, it was likely to be indirectly discriminatory. The agreement was only finally amended in 2002, after the women's committee challenged it internally.
6. See Report of the GPMU Women's Conferences 1991, 1992, 1994, 1996, 1998, 2000 and 2002.
7. It has become more common to refer to the relevant union officials as paid officials rather than full-time officials (FTOs), in order to recognise that the full-time model of union office may disadvantage women and there is no reason to insist on full-time hours. However, it is used knowingly in this book to attest to the fact that such officials were, in fact, full-time.
8. PPITB was the Printing and Publishing Industries Training Board abolished along with the other ITBs in 1981.
9. A Recruitment, Training and Retraining Agreement was concluded between the unions and the BPIF in 1983 that acknowledged the ending of apprenticeships and introduced a non-time-served modular method of training, in theory, making it easier for women to get craft training.
10. The figures used here are for manual workers in printing and publishing which therefore includes workers in publishing who are not included in the survey work for this thesis. ONS explained that this in itself is a low level of searching that may provide some erratic figures. This was certainly the case for the figures for the 1980s. However, more recently even more problems are apparent in the transfer from the New Earnings Survey to the Annual Survey of Hours and Earnings. In these latter surveys, data are no longer disaggregated for manual and non-manual workers, and for reasons that the ONS have been unable to explain, the

2002 figures suggest that women earn 80.1 per cent of men's weekly earnings and 85.5 per cent of their hourly earnings, a considerable closure of the pay gap that cannot be explained. All figures are based on median earnings.
11. As discussed in Chap. 5 experience has shown reluctance by members to reveal actual pay received. In addition a typographical error in the pay questions has affected comparability of pay at the top end but is still sufficient to indicate that differentials are far wider than in the agreement.

References

Acker, J. (1998). The Future of 'Gender and Organizations': Connections and Boundaries. *Gender, Work and Organization, 5*(4), 195–206.
Barrett, M. (1980). *Women's Oppression Today*. London: Verso.
Briskin, L. (2006). Victimisation and Agency: The Social Construction of Union Women's Leadership. *Industrial Relations Journal, 37*(4), 359–378.
Briskin, L. (2011). Union Renewal, Postheroic Leadership, and Women's Organizing: Crossing Discourses, Reframing Debates. *Labor Studies Journal, 36*(40), 508–537.
Child, J. (1967). *Industrial Relations in the British Printing Industry*. London: George Allen & Unwin.
Cockburn, C. (1983). *Brothers: Male Dominance and Technological Change*. London: Pluto Press.
Cockburn, C. (1994). The Relations of Technology: What Implications for Theories of Sex and Class? In R. Crompton & M. Mann (Eds.), *Gender and Stratification*. Cambridge: Polity Press.
Colgan, F., & Ledwith, S. (2002). Gender, Diversity and Mobilisation in UK Trade Unions. In F. Colgan & S. Ledwith (Eds.), *Gender, Diversity and Trade Unions: International Perspectives*. London: Routledge.
Colling, T., & Dickens, L. (1989). *Equality Bargaining – Why Not?* London: HMSO.
Colling, T., & Dickens, L. (2001). Gender Equality and Trade Unions: A New Basis for Mobilisation? In M. Noon & E. Ogbonna (Eds.), *Equality, Diversity and Disadvantage*. Basingstoke: Palgrave.
Craig, C., Garnsey, E., & Rubery, J. (1984). *Payment Structures and Smaller Firms: Women's Employment in Segmented Labour Markets*. Research Paper No. 48. London: Department of Employment.

Crompton, R. (1997). *Women & Work in Modern Britain*. Oxford: Oxford University Press.

Cunnison, S., & Stageman, J. (1995). *Feminizing the Unions*. Aldershot: Avebury.

Daniels, G., & McIlroy, J. (2009). *Trade Unions in a Neoliberal World: British Trade Unions Under New Labour*. London: Routledge.

Davies, H., Joshi, H., Rake, K., & Alami, R. (2000). *Women's Incomes Over the Lifetime*. London: The Stationery Office.

Dickens, L. (1997). Gender, Race and Employment Equality in Britain: Inadequate Strategies and the Role of Industrial Relations Actors. *Industrial Relations Journal, 28*(4), 282–291.

Dickens, L. (2005). Walking the Talk? Equality and Diversity in Employment. In S. Bach (Ed.), *Managing Human Resources*. Oxford: Blackwell Publishing.

Fawcett Society. (2017). *Equal Pay Day*. Retrieved November 30, 2017, from https://www.fawcettsociety.org.uk/equal-pay-day.

Findlay, P., McKinlay, A., Marks, A., & Thompson, P. (2009). Collective Bargaining and New Work Regimes: Too Important to Be Left to the Bosses. *Industrial Relations Journal, 40*(3), 235–251.

Gennard, J. (1990). *NGA: A History of the National Graphical Association*. London: Unwin Hyman.

Gennard, J., & Bain, P. (1995). *SOGAT: A History of the Society of Graphical and Allied Trades*. London: Routledge.

Graphical, Paper and Media Union. (1991). Women's Conference Report, p. 1 and Appendices A and 6.

Graphical, Paper and Media Union. (1992a). Women's Conference Report, 8, 10, 12, 13, 15, 18, 21.

Graphical, Paper and Media Union. (1992b). Women's Committee Report to the Second GPMU Women's Conference.

Graphical, Paper and Media Union. (1993). Biennial Delegate Conference Report, 78–79.

Graphical, Paper and Media Union. (1995). Biennial Delegate Conference Report, 8–9, 91–92, 99–100, 110–112.

Graphical, Paper and Media Union. (2002). GPMU Women's Committee Report to the GPMU Women's Conference.

Grimshaw, D., Ward, K. G., Rubery, J., & Beynon, H. (2001). Organisations and the Transformation of the Internal Labour Market. *Work, Employment and Society, 15*(1), 25–54.

Harris, C., & Morris, L. (1994). Households, Labour Markets and the Position of Women. In R. Crompton & M. Mann (Eds.), *Gender and Stratification*. Cambridge: Polity Press.

Heery, E., & Kelly, J. (1989). Full-Time Officers and Trade Union Recruitment. *British Journal of Industrial Relations, XXVII*(2), 196–213.

Jones, A. (1969). *Cmnd. 4219. Pay Structure Within H.M. Stationery Office Presses and Binderies*. London: HMSO.

Kirton, G., & Healy, G. (1999). Transforming Union Women: The Role of Women Union Officials in Union Renewal. *Industrial Relations Journal, 30*(1), 31–45.

Ledwith, S. (1991). *Sweet Talk, Soft Deals? – Gender and Workplace Industrial Relations*. Working Papers in Economics, Business and Management No. 35. London: Polytechnic of North London.

Lukes, S. (2005). *Power: A Radical View*. London: Palgrave Macmillan.

Musson, A. E. (1954). *The Typographical Association*. London: Oxford University Press.

National Graphical Association. (1976). Biennial Delegate Meeting Report, p. 24.

National Union of Printing, Bookbinding and Paper Workers (NUPB&PW). (1953). Circular 17th May.

New Earnings Survey. (2002). By Industry.

Nicholson, J. (1994). *The Influence of the London Women's Branch of the Printing Unions on Women's Employment in the Industry: Demarcation, Inter-Union Strife and Employment Restriction 1950–1960*. Unpublished Dissertation, London College of Printing.

Noon, M. (2007). The Fatal Flaws of Diversity and the Business Case for Ethnic Minorities. *Work, Employment and Society, 21*(4), 773–784.

Office for National Statistics (ONS). (2014). Annual Survey of Hours and Earnings by Industry – Tables 16.1a Gross Weekly Earnings, Not Affected by Absence and 16.5a Gross Hourly Earnings, Not Affected by Absence.

Office for National Statistics (ONS). (2017a). Table 13, Employment by Industry.

Office for National Statistics (ONS). (2017b). Tables 14.1a and 14.5a 4-Digit SOC.

Office for National Statistics (ONS). (2017c). Tables 16.1a and 16.5a 4-Digit SOC.

Phelps Brown, H. (1977). *The Inequality of Pay*. Oxford: Oxford University Press.

Printing and Publishing Industry Training Board (PPITB). (1982). *Equalising Job Opportunities for Women in Printing and Publishing: An Attitude Survey Report*. London: Printing and Publishing Industry Training Board.
Roe, A. (2003). *The National Agreement in the Printing Industry – The Exception to Test the Rule*. PhD Thesis Unpublished, Keele University, Staffs.
Rubery, J. (1997). Wages and the Labour Market. *British Journal of Industrial Relations, 35*(3), 337–362.
Rubery, J., & Fagan, C. (1995). Gender Segregation in Societal Context. *Work, Employment and Society, 9*(2), 213–240.
Rubery, J., & Rafferty, A. (2013). Women and Recession Revisited. *Work, Employment and Society, 27*(3), 414–432.
Siltanen, J. (1994). *Locating Gender: Occupational Segregation, Wages, and Domestic Responsibility*. London: UCL Press.
Smith, P., & Morton, G. (2006). Nine Years of New Labour: Neoliberalism and Workers' Rights. *British Journal of Industrial Relations, 44*(3), 401–420.
Society of Graphical and Allied Trades. (1955). Report to Executive Council, Modern Records Centre, Warwick University (MSS39A/TA/1/121).
Society of Graphical and Allied Trade. (1990, April). *SOGAT Journal*.
Society of Graphical and Allied Trades. (1974a, May). *SOGAT Journal*, pp. 6–7.
Society of Graphical and Allied Trades. (1974b, July). *SOGAT Journal*, pp. 2–3.
Society of Graphical and Allied Trades. (1974c, July). *SOGAT Journal*, p. 17.
Society of Graphical and Allied Trades. (1975a, March). *SOGAT Journal*, p. 12.
Society of Graphical and Allied Trades. (1975b, May). *SOGAT Journal*, p. 6.
Society of Graphical and Allied Trades. (1980). SOGAT Journal, June, p. 15.
Walton, R., & McKersie, R. (1965). *A Behavioral Theory of Labor Negotiations*. New York: McGraw Hill.
Willis, P. (1977). *Learning to Labour*. Farnborough: Saxon House.
Wootton, B. (1955). *The Social Foundations of Wage Policy*. London: George Allen & Unwin.

8

'Paid Class 3, Treated Class 3, Act Class 3[?]': Overcoming the Barriers

The position of the author as equality policy adviser (EPA) for the Graphical, Paper and Media Union (GPMU) elicited the three questions that inform this study into the British printing industry. What are the causes of the gender pay gap (GPG) in the printing industry? Why has the gender pay gap been so persistent, despite technological change? Are there conditions under which the gender pay gap might change? The research was restricted to the general printing industry and used historical analysis and a sample survey of men and women trade union members employed in the general printing industry, interviews with employers, trade union officials (nationally and locally) and male and female union members. What emerged was the importance of access to power resources if women are to successfully challenge embedded discrimination.

Chapter 2 addressed theoretical approaches to explaining discrimination. Given that the research sought to uncover both class and gender effects on pay, dual systems suggested itself as a starting point. However, its notion of two separate systems in articulation has been regarded as inadequate for being too functional for capital, and ignoring tensions between the systems (Walby 1986), for tending towards ahistoricism and essentialism (Acker 1998) and because it is not clear how patriarchy and capitalism interact or what makes them systematic (Connell 1987).

T. Dawson, *Gender, Class and Power*, https://doi.org/10.1057/978-1-137-58594-3_8

Instead, utilising a looser conception of intersectionality (Crenshaw 1991) and the analytical tool of inequality regimes (Acker 2006) has provided firmer ground on which to unpack discrimination. Understanding the unequal access to power resources allows for a subtler account of change over time in the printing industry and how women fought for influence but were generally outflanked by employers and male workers. Bradley's approach to understanding power applies to class and gender a concept of the dialectic that operates between 'material' and 'cultural' aspects (Bradley 1999: 21–23) that bridge the gap between structure and action (1996: 5–7). This analysis of the printing industry has adapted Bradley's approach and taken it a step further to explain the embeddedness of gender discrimination. Chapter 2 explains this model, its extension to encompass legal power and its ability to demonstrate how some power resources enable access to further power resources, which then operate a continual reinforcement mechanism. It is hard to disrupt this process, once established. It, thus, confers relative power on the parties to the employment relationship, which in turn, provides relative privilege. The result is to make power relations seem natural (Lukes 2005) to the extent that relative power and privilege become invisible to the beneficiaries. Challenges to this 'natural' order are frequently disparaged as 'identity politics' (Fraser 2008) that undermine solidarity around class. Thus, the identity of the power holders is taken for granted and seen as unproblematic. These processes provide an explanation of the resilience of the gender pay gap, through 'practices of *reproduction, rationalisation and resistance*' (Collinson et al. 1990: 193 original italics) in work organisation and pay processes.

Class and gender dynamics are revealed in the historical development of the printing industry, especially in work organisation and its link to occupational segregation and exclusion of women. The stability of technology in maintaining work organisation is apparent. Serious challenges to male craftworkers' control of the labour supply emerged post-1945, especially post-1980. Increasing technological change affected both the balance of power between organised workers and capital and also gender relations. Meanwhile capital's tendency to overcapacity and collapse (Cockburn 1983) reached crisis point in the early 2000s with major loss of jobs and union members in the printing industry. Once the 'tipping point' was reached, the majority of male print workers saw a drastic reduction in the basis of their power. Skill had been that source of power but became useful for a decreasing number of manual print workers. Moreover, the majority of the remaining work moved to the communications sector where union organisation remains weak and the workforce is more diverse. The failure to accept the changing nature of the printing industry and *inter alia* to relate to women workers as equals has prevented the union from building cross-gender solidarity and 'following the job' into growth areas. The global financial crash (GFC) exacerbated this process, and only the strongest chapels have subsequently started to regain lost terms and conditions in 2017. While gender relations were not the only important factor in the GPMU's demise, internal organisational weakness also being a key factor, they were certainly significant.

The result, for women workers, was uneven, sometimes one step forward, two steps back but latterly vice versa, women actually losing their place in the industry and the limited protection of the secondary internal labour market (SILM) they inhabited. Their invisibility reduced for a time but subsequently increased significantly. This chapter therefore considers the processes leading to women's invisibility in the workplace and in bargaining processes. The interplay of structure and agency reveals men's and manager's continued efforts to reproduce and rationalise barriers to women's employment in 'men's work'; their resistance to changes in paid labour; and the complexity of women's resistance to unequal treatment in work, the trade union and bargaining processes, alongside their strategies of compliance. Women's struggle for equality, in this sector, has continued in a context of economic restructuring, weaker trade union organisation and more individualised bargaining processes—where they have maintained a presence they have largely continued to be 'paid Class 3, treated Class 3'—the extent to which they have 'acted Class 3' (R2[4]) will form part of the following discussion.

Women's Invisibility in the Work Organisation

This section considers processes of occupational segregation, their elision with labour market segmentation and how this affected gendered power relations. Union and employer battles over the labour process are central to this discussion. The effect of domestic labour on women's access to employment is also explained, briefly. These processes demonstrate the shifting pattern of women's invisibility over time and argue that this helps explain the maintenance of the gender pay gap.

Chapter 3 traces the history of printing in Britain up to 1945. The division of craft labour was established and maintained from the introduction of printing to Britain in the late fifteenth century (Delafons 1965). The economic power of the guild regulated pay, prices and, to some degree, labour at a politically sensitive time. The masters gained the benefits of guaranteed prices (Child 1967) and cheap labour from abusing the apprenticeship system (Delafons 1965). Patriarchal gender relations provided technical and collective and (relative) economic power for male workers, with only limited encroachment from masters' wives. At this stage women were literally invisible.

From the late eighteenth century onwards, a more fundamental division of labour between pre-press and printing occurred, splitting a key element of control between craftsmen and 'crude mechanics' (Musson 1954: 51). Women remained barred from craftwork, and initially there was male resistance to women being employed in the industry at all, enforcing their invisibility. So, the adaptation of guild practices to capitalist production relations strengthened men's hold over work organisation and women (Child 1967; Hufton 1995). When women did enter the industry in the second half of the nineteenth century, there were a number of disputes between employers and men that were significant in establishing the extent of male resistance to female encroachment (Musson 1954; Bundock 1958; Gillespie 1953; Reynolds 1989). These disputes developed a clear demarcation between what was regarded as skilled and male and what could be designated as work suitable for women. Women resisted where possible but were either forced to concede (Drake 1920) or chose to comply for reasons of class solidarity (Reynolds 1989). In general, though, women did not threaten men's jobs as they were directed into occupations created by a further division in the labour process in the binderies (Duffy 2000). Men had strengthened their technical, collective and, therefore, economic power resources to ensure, contrary to Walby (1986), that both exclusion and occupational segregation occurred simultaneously in the industry.

It is argued that skill acted as a proxy for class relations in print—it was the arena in which men and employers sought to gain advantage. It acted as a hidden source of economic and symbolic power for employers in ensuring that the jobs women undertook were low paid; skilled work was something women did not do (Phillips and Taylor 1980). For men, skill was a source of technical power that created and sustained male advantage, ensuring that they filled higher places in the hierarchy than women (Hartmann 1981) and protected their jobs through restricting the labour supply. From the beginning, the identification of masculinity with physical power and skill played a role in ensuring women remained in segregated jobs. Women's occasional encroachment into skilled work left men potentially only able to claim sovereignty over heavy lifting. The contradiction, that heavy lifting is not a skill but a physical attribute, seems to have been conveniently overlooked in the battle for job control (Cockburn 1983). Furthermore, while women are often criticised for focusing on their gender identity (e.g., Fraser 2008), men's fiercely protected identity is so embedded it is invisible and, therefore, not seen as problematic.

Subsequently, the sexual division of labour became institutionalised because of the relative stability of the labour process (Zeitlin 1985). Even the potential threat posed by women dilutees in the two world wars was successfully controlled (Musson 1954; Howe 1950; Child 1967). Employment policies reinforced the gender inequality regime. Recruitment, training and promotion decisions were, and generally remain, largely in the hands of time-served line managers who have tended to perpetuate the pattern of employment, as evidenced in Chap. 6. In particular their reliance on informal networks for recruitment and what was identified as a 'family effect', in Chap. 5, tended to replicate traditional gender roles in the industry. The effect of equal opportunities policies, it is argued here, has been to proceduralise discrimination, affording formal equality but not substantive equality. Meanwhile, men's shared symbolic power, defining work appropriate to men and women, reinforced exclusion and segregation. This elided with labour market segmentation; women were not considered for skilled jobs, and it is likely that anticipated discrimination ensured few women tried to escape this designation.

In line with Lukes' contention that hidden power can 'naturalise' processes (Lukes 2005: 28), the asymmetrical power relations that resulted in occupational segregation rendered women largely invisible and rationalised women's retention in low-paid jobs. There were, however, two important phases of encroachment, where women resisted this designation. The first in general printing hardly displaced men and was temporary; the second, in newspapers, was linked to large-scale displacement of men and union derecognition (see Chaps. 5 and 7). Subsequently, employers utilised a familiar capitalist strategy (Hacker 1990) and disposed of women, too, it would seem, at least in part, in favour of cheaper migrant and/or agency labour. The women's finishing work that remained was often outsourced. Women, ultimately, became almost literally invisible again.

There is an irony in the gendered focus of male print workers, seeing women as the greatest threat to their wages and conditions. The most dangerous segments were always other men because they had better access to skill formation. In the nineteenth century, the greatest vitriol was reserved for women strikebreakers, even though it was often men who had the most significant effect on employers' ability to maintain production (Reynolds 1989). Similarly, the advent of a new production worker, the machine minder, led to a breach in craftworkers' opposition to 'progression'. The same was often true of 'boy' labour. It was only women who were not allowed 'progression' (see Chaps. 3 and 4). In so far as women succeeded in gaining skilled work, re-segmentation ensured they could still be differentiated from men, to their detriment, as in the short-lived access to typesetting jobs in general print.

The continued resistance to female labour, it is argued here, blind-sided male print workers, who continued to ignore the shift to communications and disappearance of work abroad. As such, the truly progressive potential merger across the craft divide, in the 1960s, failed, and it was almost 30 years before it came to pass, by which time it was, apparently, too late. They also ignored employers carving up Class 1 jobs, so that ever-smaller proportions of men could continue to be designated as high skilled. They were subsequently too weak to resist the use of agency labour, often migrant labour in the 2000s, yet another form of segmentation.

Whereas latterly segmentation served employers well, in the early phases of industrialisation, it also restricted their strategies. As Edwards (1995) pointed out, managers need to develop a balance between conflict and co-operation. This acts as a brake upon the divide-and-rule strategy, as do ILMs to some extent. Given craftsmen's domination of the workforce and strong union organisation, this could explain the reluctance to 'take on' the unions prior to 1945. The growing confidence of employers, especially from the late 1990s onwards, has led to greater segmentation and a focus on performance. The later research, however, suggests that the need for co-operation may be increasing again, as male print workers in strong chapels (workplace branches) begin to claw back some of the terms and conditions they forfeited in the wake of the GFC.

This study has not been able to delve deeply into the role of domestic labour in restricting women's employment in printing. What was clear, from the survey, was the extent to which domestic gender roles had been preserved, with women retaining the lion's share of responsibility for domestic labour and caring. Furthermore, despite the waning of the father-to-son route into the industry, a 'family effect', whereby sons were directed into printing and daughters into the bindery, had clearly affected gender composition of the workforce up to 2002. There was one telling result, however, which was that both the survey and interviews with women members suggested an element of self-selection for work in the industry among them. The survey demonstrated that a relatively small proportion of respondents had childcare responsibilities, while interviews suggested that women would enter the industry for several years, leave when their families were young and then return at a point when childcare was manageable. This seems to be a function of employer's reluctance to work 'round those with kiddies'. But the effect of the secondary ILM probably made it worthwhile for women to return to the industry as soon as they were able. So, not only did women lose out in developing power resources by late entry to the industry, but employers' gender inequality regimes ensured that domestic responsibilities continued to disadvantage women throughout their working lives.

Crucial to understanding how gendered work organisation emerged is the role of trade unions. Print unions organised according to a plethora of craft distinctions based on skill definitions, incorporating masculine traits, among them mental and physical toughness, competitiveness and aggressiveness (Cunnison and Stageman 1995). As such, the elision of technical with sexual and collective power resources resulted in a strong gender identity that, when in conflict with management's economic and positional resources, became a proxy for class relations. This justified the marginalisation of women workers and their concerns, in the name of class solidarity. However, even class solidarity was limited by the sectional interests of the different unions, having its most beneficial effect (for male print workers) in the leapfrogging pay claims of the 1950s (Gennard 1990).

In this regard, the strategy of excluding women and segregating them when exclusion was impossible makes sense, as does the organising of women into their own unions when this strategy failed. Women were apparently invisible and irrelevant to the industry—or that's how men would have preferred it to be, not because of overt misogyny but because gendered attitudes made this seem a 'common-sense' approach: clearly women were a threat and so keeping them out of the way was the safest option. However, this never worked in practice, allowing employers to create new divisions in work organisation and utilise women's cheap labour. The gender divide apparently suited employers.

Employers found the issue of women's organisation more complex. Their immediate concern in the late nineteenth and early twentieth centuries was to develop scientific costing in order to ensure business viability. In these circumstances, however counterintuitive it may seem, encouraging women's union organisation would have assisted these employers in suppressing undercutting (McDonald 1904). Their economic power was under threat so they were minded to accept what was normally unacceptable.

For women, they were caught between their class interests and their gender interests because unions at that time did not recognise the importance of specifically female class interests. When given the opportunity to organise, they proved themselves capable but limited by labour market segmentation (Bundock 1958). They seemed to be incorporated within trade unionism more on sufferance until after the Second World War—controlled during both world wars by recourse to male trade unionists technical and collective powers, which gave them access to legal power, that is, the government acquiesced to women's removal from 'men's work' as soon as the wars ended (Musson 1954). However, from the 1960s onwards women began to realise a greater degree of collective power, campaigning for equal rights and gaining a measure of legal power as a result. Trade unions reflected this social change in the TUC Charter *Equality for Women Within Trade Unions* (Gennard and Bain 1995). Peer pressure led to internal structural change within print unions, alongside pressures to merge, with labour process developments emanating from technical change. Merger meant less sectionalism, although the skilled/unskilled divide was, arguably, never really overcome, even after the merger to form the GPMU.

Considerable structural change only arrived with the development of one union for the industry. Guaranteed proportional representation (GPR) gave women places on decision-making bodies from 1995 but only in proportion to their numbers. As women were losing jobs and the merger had diluted their numbers, they only amounted to 17 per cent of the union, insufficient to form a critical mass that could lead to transformational change. Their access to jobs was still constrained, limiting their access to technical power, putting a brake on their ability to leverage greater collective power. The reinforcement mechanism, whereby power resources could multiply, was therefore weakened. While they undoubtedly gained greater legal power, their personal and symbolic power resources never really grew significantly. Male trade unionists had more or less monopolised personal power; women struggled to learn the 'rules of the game' and were, anyway, reluctant to engage in the kind of internal politicking that could challenge existing assumptions about allocation of jobs and union positions (Colgan and Ledwith 1996). GPR was easily undermined as powerful branch secretaries learned to game the system. It flowered briefly, demonstrating women could, in some circumstances, get elected in their own right, but it was withering away by the time of the Amicus merger. Even when women acquired senior positions, as with Brenda Dean, a lone woman was constrained in what she could achieve against such male domination—and she seemed reluctant to be drawn too deeply into gender politics (Dean 2007). When it came to women in full-time officer positions, they were engaged in an uphill struggle to be taken seriously as negotiators (Willis 1977). In essence, internal structural change was a form of proceduralisation that provided formal, but not substantive, equality and was easily marginalised once men adapted to the new 'rules of the game'.

These practices may help explain women's apparent collusion with men in their subordination—giving up lucrative work when men called on them to do so. In reality, they were caught between their class loyalties and their gender interests. Men's gender interests were taken as the norm and as natural, it was unthinkable that this should be challenged (Lukes 2005). Even women who worked in the industry just after the war still accorded primacy of place to time-served men (see Chap. 4). When taken alongside the secondary ILM that women print workers inhabited, the relative economic power they acquired compared to what was available in other 'women's work' was probably sufficient to encourage them, not so much in collusion, as rational compliance.

Overall, men's access to technical power and their guild experiences encouraged the formation of collective power in the face of capitalist development. This encouraged a strong masculine identity and it concomitant sexual power. Thus, they acquired great job-related skills and knowledge, which delivered recognitional domination (Lukes 2005), alongside positional, personal and symbolic power within their unions. Consequently, they developed strong controls over skill definitions and the intra-union 'rules of the game'. At key points, they even accessed legal power, as in the two world wars. Such overwhelming access to power resources made it easy to outflank women workers. While this must be seen in the light of class relations driving men to feel the need to develop a strong position if they were to resist employers, it had long-term, adverse consequences.

By the time women acquired sufficient power to begin to challenge these practices, the industry was beginning to restructure. Women lost their jobs faster than men, and the men did not follow the job into more diverse sectors. The contraction in the industry has continued to favour men. However, men have decreasing opportunities to access the highest-paid jobs, which have diminished. Men's jobs have been segmented, often in favour of agency labour, significant proportions from migrant communities. The failure to overcome sectional interests at an earlier stage set the scene for diminishing union strength as union's legal power reduced in favour of employers. Employers economic power has soared in the age of neoliberalism. Gender relations ultimately undermined class relations—not because of women's use as cheap labour but because men refused to sully their status by sharing it with women.

Women's Invisibility in Bargaining Processes

The main reason why women's invisibility might have led to a perception of their irrelevance to bargaining processes was in the strategy of wage leadership (Cockburn 1983: 201). As such the focus was fixed only on the key group in bargaining processes because everyone else's pay depended on a good settlement for that group. There was also a contradiction here. Cockburn suggested resentment towards 'the lower ranks for "climbing on the skilled man's back" in pay negotiations' (ibid.: 128), although this attitude was not apparent in this research, which might suggest support for women was growing.

As explained in Chap. 7, the national agreement institutionalised basic pay relativities. At that level pay differentials appeared to be very narrow. In reality, the implementation of the agreement meant that craftsmen got a great many lucrative extra payments that significantly inflated gross pay differentials. The women in the binderies only got extras if they did overtime, for example, but far fewer got shift pay or the other extras the men received. Women generally received only relatively small enhancements to the basic rate.

While women tended to get paid considerably in excess of the national minimum wage, the weakness of the union saw women's pay fall towards this level; any union premium on women's wages was eroded as men's technical and collective power diminished. Employers were quick to seize the initiative, driving down all wages once they had the power to take on Class 1 workers.

Until recently, women were largely excluded from the bargaining table (see Chap. 7). Negotiators tended to be male, craft-trained trade unionists. The employers' federation panel in national negotiations was almost always male, consisting of a combination of professional federation staff of long-standing and employers with long experience of the industry.

As Walton and McKersie suggested (Walton and McKersie 1965: 185), years of facing each other across the bargaining table could establish a pattern of bargaining.

An important method of forcing women's issues back into invisibility and preventing transformation of the bargaining agenda is captured by the concept of delegitimation. This affected, on the one hand, issues of importance for women and, on the other hand, the identity and standing of negotiators. The employers' interviews suggested a fairly cynical approach to equality bargaining, delegitimising equality as inappropriate 'social issues' on the one hand and seeing the union as merely submitting a 'business case' for equality bargaining on the other (Dickens 1997). In such circumstances, they did not expect sustained pressure from the union on equality issues, and they rarely got it, indicating that union officials largely accepted the designation of equality issues as 'social'. As Chap. 7 demonstrated, women officers rarely came from the shopfloor, and this was used to delegitimise their suitability as negotiators; such women were not seen as identifying with production workers or having sufficient experience of shopfloor production relations. Although women are often accused of undermining class solidarity through identity politics (Fraser 2008), this example would seem to privilege male identity at the expense of women gaining positional power. Male identity trumped all in this situation.

While this study did not consider the role of domestic labour in depth, evidence certainly demonstrates that women retained primary responsibility for this work. This would serve to embed the perception (and definition) of women's roles as that of secondary workers: a form of hidden symbolic power (Lukes 2005; Bradley 1999), thus limiting women's self-determination.

While minorities are not necessarily doomed to be ignored in bargaining processes, their lack of a critical mass can sustain pay inequality. Women had accounted for higher percentages in SOGAT that had encouraged some moves towards closing differentials but no challenge to occupational segregation (see Chap. 4). Former NGA interviewees were quick to point this out. However, one telling quote from officer 13 also provided a possible reason for why GPR was conceded; a low proportion of women's membership meant it would not cause too much disruption and therefore 'we got it about right'. Concessions, such as they were, only occurred when the union had been weakened, as with other unions (Colgan and Ledwith 2002).

For much of the print unions' history women periodically proved useful, despite their lowly position at work and in pay grades. Employers have, when wishing to try and gain some financial advantage, initially targeted women. As examples in Chaps. 3, 4, and 7 revealed, in these instances men defended women's earnings but only within the context, ultimately, of defending their own. So, women's pay was often the battleground over which men as employers and workers sought to pursue class antagonisms. This may have been justified in the union's mind as protecting the male breadwinner's wage, supporting the family, but it was at women's expense and ensured their earnings were viewed only as component wages (Siltanen 1994), as witnessed in the Format International dispute (see Chap. 4).

In the 1990s women's differentials became a potential battleground again, but this time there was little fighting. True, the union was weaker, but for several years there had been a growing feeling among some senior union officers that differentials were too narrow, a view originally voiced among newspaper printers in the 1980s (Cockburn 1983; Chap. 4). Employer 1 acknowledged that it was difficult for the union to openly accept widening of differentials, so the parties apparently agreed it would only be temporary but it was never closed again or even tabled for discussion. Women, generally, were not directly consulted about the effect of this agreement on their earnings, although male-dominated branch committees were. While this may constitute collusion, union weakness probably underpinned their contradictory attitude, occurring as it did around the time of the discussions to introduce GPR.

The irony of this episode was that it did little to protect male craftworkers, although it reinforced the gender hierarchy. Unlike their predecessors, who had seen the benefit of using women's pay to protect all workers' rates of pay (see Chaps. 3 and 4), women were sacrificed in order to apparently gain increased funds for employers to continue poaching men in a situation of skills' shortages—when skilled workers constituted 70 per cent of the workforce. This was made clear by employer 1 in Chap. 7. It also allowed chapels that remained strong to improve on the low rises achieved in the national agreement that were causing widespread discontent. However, within a couple of years, pressure was being exerted on men's rates too. Companies increasingly failed to pay the national increase, and by the end of the decade it was pay cuts that were the subject of many negotiations, not rises (see Chaps. 4 and 7). Increasingly, employers were able to impose pay increases with no union involvement.

The business case for equality also failed to serve women's interests. As employer 1 indicated, production requirements will always take precedence over diversity. Therefore, while the business case might provide a bargaining tool, the social justice case cannot be ignored if equal pay is ever to be a reality (Noon 2007).

It was the inherent contradiction of men's views on women print workers that reinforced these tendencies. On the one hand, there was a view expressed, whenever women broke through the exclusionary barriers, that they could not do the job so were not worth the rate, but also a fear that they were cheap labour. Men were trapped between the undermining of their masculine identity as skilled printers and the fear of being undercut by women quite capable of doing the job, challenging men's status. This prevented men from ever forging a real collective identity with women to resist employers' desire to cut pay and intensify work. This may be seen as a reluctance to give up collective, sexual and technical power. It also constituted a success for employers' divide-and-rule strategies, but one that the men blindly conceded.

Unionised women did not remain quiescent and achieved a greater voice on the executive council (EC) through GPR. Subsequently, one of the EC women was included on the national negotiating panel. The women's committee challenged their exclusion from discussing aspects of the national agreement and continued to recommend policies that widened the bargaining agenda. However, little was achieved in the agreement beyond restatements of legal rights or statements of intent (Ledwith 1991). Designating women's issues as 'social issues' also hid their material effect on pay but went unchallenged, reflecting a consensus between male negotiators and employers on what constituted 'industrial' issues, an example of symbolic power.

Therefore, women's long battle to get items of importance to them on the agenda was only the first step. To secure actual benefits required tenacious bargaining, but, in reality, these items tended to come up early in discussion, quickly reach an impasse and then disappear until they reappeared in the next annual round of bargaining—a form of proceduralisation that gave an appearance of activity but few concrete results. However, there was one glaring omission from the agenda—barely any discussion of the grading system took place since SOGAT sought to remove Class 3 from the agreement in the 1980s (see Chaps. 4 and 7). Instead there was a forced focus on women's access to jobs—occupational segregation was the problem, not the valuation of women's skills, despite successful legal challenges on this issue (see Chap. 7).

Union officials highlighted the difficulty of getting men to support negotiations to secure benefits that were primarily for women, although not all agreed (see Chap. 7). This meant that, nationally, there was little bargaining power for negotiators seeking improvements for women. But neither was there a campaign involving women members to try and win men's support. Add to this the admission from interviewees that they did not really police the national agreement, and even the inadequate legal rights contained in it were rarely enforced, except as a result of individual grievances.

Union weakness resulted in greater discontent among all members, with the low settlements achieved under the national agreement during this period, and the scaling back of second-tier bargaining providing few opportunities for improvements. Traditionally, second-tier bargaining had provided opportunities, for men especially, to improve their pay further. The gender divide was most apparent here, as departmental bargaining reflected the division of labour and so men could bargain for improvements without women gaining any benefit (Ledwith 1991). However, because skill was no longer such a useful bargaining chip, it may be assumed that there was more scope for breaking down the gender divide. There is no evidence for this happening but more a desire to protect what each member held. As a number of members stated, 'there is a look after number one attitude now' (see Chap. 7). At this local level women were generally absent from negotiations and unable to influence them.

The issue of skill is complex in this context. For many years there were skills' shortages, but employers relied on poaching trained workers, rather than invest in training. At the same time technology stripped away skill from successive layers of male workers. Skill remained useful for only a minority, mainly No. 1 printers. These contradictory processes never gave access for women to skilled work, but, in terms of bargaining processes, union officers could continue to use lack of access to skilled jobs as the reason for women's low pay—not a faulty evaluation of 'women's work'. This may have informed resistance to women's legal power in the case of J & C Moore, where the interpretation of the agreement was seriously challenged. Coincidence of interest between employers and the union prevented this legal decision having wider application.

The effect of these bargaining processes continued to ensure women remained invisible. Men's positional power in the union, reinforced by their personal and collective power, generally led to a domination of both the agenda and its negotiation, whereby women were perceived as irrelevant to bargaining processes. Crucial support came from symbolic power, allowing employers and the union to delegitimise women's bargaining issues. The coincidence of interests between employers and male trade unionists and unequal access to power resources exacerbated reinforcement of gendered power relations. While women's sexual and collective power provided some scope for resistance, it was insufficient to affect outcomes. Women were small in number on influential committees and negotiating panels and at local level were generally absent altogether. Women needed greater positional and collective power to challenge the operation of the grading structure and implementation of the national agreement in order to achieve a fairer valuation of their contribution to production. The individualistic nature of their legal power was insufficient to create real and lasting change. Women were, in the end, outflanked by those who already had greater power resources.

Summary and Final Comments

This study adopts an industrial relations approach but with the added dimension of a gendered analysis of workplace processes and practices. It identifies the intersectionality of class and gender relations. In industrial relations class has often been taken as an unproblematic lens through which to view the workplace. Conversely, this study identifies how men's identity has been central to how they pursued class antagonisms and how recourse to class was used in an attempt to delegitimise gender analysis for undermining class solidarity. Men misdirected attention from their gender identity, rendering it invisible in much industrial relations analysis, and problematized women's identity as antagonistic to class relations. This dovetailed with employers' interests in accessing women as a source of cheap labour. Although collusion undoubtedly occurred, as both the historic account and Ledwith (1991) demonstrated, it has been shown here that this is not a necessary condition for gender inequality to thrive; coincidence of interest is sufficient.

Real dangers arose from this approach. It distracted male trade unionists from recognising the dangers, in favour of maintaining existing gender relations. Effectively, men took their 'eye off the ball' as employers segmented the workforce in new ways, with the result that fewer workers were able to access the highest pay, including many men. This process was exacerbated by the use of government power resources (not specifically covered in the model in Chap. 2) to favour employers and not unions. Managerial prerogative was strengthened from 1979 onwards, re-regulating unions, while de-regulating employment, the economy and finance (Fosh et al. 1993; Thompson 2011), coercive government power, giving way to New Labour's non-decision-making, largely supporting the status quo in industrial relations (Smith and Morton 2006; Daniels and McIlroy 2009).

The contention of this study is that using a model of power resources can reveal barriers to achieving equality in jobs and pay. In theory, women in print had gained influence in their union through their access to, and greater confidence in using, collective and sexual power resources. They were even starting to achieve access to secondary power resources, personal and positional power and, just possibly, a little symbolic power, allowing them to challenge hegemonic understanding of the workplace and bargaining agenda. What this analysis demonstrates is how structural change in the union was insufficient to deliver transformational change. Men's greater, and more embedded, power resources acted as a brake on progress, especially when it became clear that GPR could deny some men seats at the top table. When confident enough to stand for election, women were shown to be popular candidates. So, the familiar pattern of 'two steps forward, one step back' was apparent. But by the third round of GPR for the EC, women could not get enough seats to fulfil proportionality. At the same time, they had been losing jobs faster than men. Arguably, it had become more a case of 'one step forward, two steps back'.

So, the case for the effect of power resources is quite clear—they can act as a brake, or even a complete block on women's progress towards equality. But what does this mean beyond the printing industry? It is argued here that the power resources model is applicable to all industries and sectors. While the picture in each case will not be identical, its broad parameters are likely to be similar. For example, while non-manual work may not include a central role for physical power in the sense of heavy lifting, given recent revelations (see the Harvey Weinstein debacle, 2017), it is possible that physical power plays a role in sexual harassment—physical and sexual power intersecting. Even more forcefully, it is likely that any sector will have skill or knowledge requirements. Thus, technical power is likely to prove central to gender relations in all sectors. In particular, the social construction of skill (symbolic power) and, in manual working, its conflation with physical power gives rise to a high probability of gendered work organisation. Munro (1999) identified this process in her study of the NHS, whereby even cleaning equipment acquired a gender identity. When bigger, more powerful floor cleaners were introduced, male cleaners laid claim to them and designated their own hierarchy, where none existed previously.

When applying the model to pay processes, both the agenda and the identity of negotiators would require examination. Where there is a GPG, there is likely to be a process of delegitimation in action, operating through symbolic power. Clearly, in the NHS, Agenda for Change was introduced, at least in part, to address gendered pay structures. Equal pay cases in local government and the NHS (BBC 2013; The Guardian 2010) demonstrate how pay processes favoured men. Legal power resources were required to successfully challenge these. In such cases, union officials found themselves caught between class and gender imperatives. Arguably, a greater understanding of the processes examined in this study might have assisted them in negotiating such tensions. Instead they were cruelly exposed in the media as being complicit in denying many of these women the full settlement they were entitle to, despite having secured equal pay for the present and future. They appeared to be subject to greater opprobrium than the employers. In this case, the government largely sidestepped its responsibility by refusing to fund the removal of the GPG.

Once again, this raises the issue of government power resources, greater than for all the other parties to the employment relationship, though not specifically addressed in the model. It is government that affects the balance of power in industrial relations, choosing to favour employers to a much greater extent than unions since 1979. It has enhanced managerial prerogative by weakening unions, thus management have excessive positional and economic power and no need to address gender equality, except when public opinion calls for it (a function of women's collective and sexual power resources in campaigning). Employers' personal power in influencing government policy has also been greatly increased during this period.

The model of power resources used to examine work organisation and pay processes in the printing industry is predicated on access to primary resources, which once acquired, give access to secondary power resources. These two categories of power resources then operate in a continual reinforcement process, unless a significant rupture in industrial relations occurs. Looking at each of the parties, this process can be, briefly, illustrated.

In essence, employers acquire decisive amounts of economic power by virtue of controlling organisational resources (positional power) and benefiting from capital accumulation. They may access this power by claiming expertise as managers (Storey 1983), technical power. Or they may inherit wealth, meaning they would enter the model in the middle, a different route to other parties to the employment relationship. They may acquire collective power too, as in membership of an employers' association. As argued below, they may exercise personal power to gain legal power, by influencing government. Finally, they utilise symbolic power to define work roles, including their own, the bargaining agenda and its context (such as economic forces affecting the business). Once they have access to either technical and positional power resources or enter directly through access to economic power resources, they gain access *inter alia* to the invaluable personal and symbolic power resources that begin the reinforcement mechanism. Government plays a decisive role in the extent of managerial privilege conferred as they choose to what extent they support the parties to the employment relationship.

Male workers rely, essentially, on technical power, which they leverage through sexual and technical power resources, and to some extent physical power too. This confers relative economic power and access to personal, positional and symbolic power through union organisation. The reinforcement process embeds their relative power in the industry. For those in less skilled jobs, symbolic power would be very powerful in controlling relative skill hierarchies. The late entry of women, conversely, limits the extent of their power resources. Without initial power, they cannot easily dislodge technical, sexual and collective power nor the symbolic power that defines their very existence in the labour process. They are always playing catch-up. Their most effective power resources have been their collective and sexual power resources, utilised to campaign for legal power. But the result has been to acquire formal equality through prodecuralisation within organisations' inequality regimes, in gendered negotiating processes, or within unions' gendered representation processes. Employers' economic power has been decisive in removing their jobs from the industry, undermining any equality initiatives, rendering women and their issues relatively, if not completely, invisible again. Such is the power imbalance that employers and, until recent times, unions could act in studied ignorance of gender inequalities. Employers and male workers could afford to be oblivious (until recently) of the embedded privilege conferred by the gendered asymmetry of power.

In these circumstances, it is hardly surprising that women struggled to access sufficient power resources to successfully challenge employers or male workers' power. It is the slow accretion of power to women that accounts for the glacial progress in gender equality. However, employer resistance has been largely decisive since 1979. Unions have progressively accepted change in gender relations but at a time when their collective power has weakened considerably. They have still achieved more for women's equality than any other institution, however, flawed they remain. Reflecting on the GPMU equality initiatives, women interviewed (between 2014 and 2017) acknowledged how far the union had travelled. Their shock at the distance between the GPMU and Amicus, and even latterly Unite, in terms of attitudes was marked. This suggests that while, initially, the minds of men in the GPMU were won over by rational argument, by the time of the merger with Amicus, some deeper transformation had begun to take route, affecting hearts too. Thus, even the former GPMU men were shocked at the attitudes towards women and minorities on display in Amicus. This also suggests the power resources model could usefully be applied to current union organisation to expose the barriers to substantive equality. As such, union structures may still only be providing formal equality. It is still, however, likely that unions will continue to be central to any gains women make in the foreseeable future.

Question 3 has not yet been addressed and asked under what conditions might the GPG change. Clearly, the reflection of activists and former officers on the GPMU, in the second period of research, highlighted how important structural changes were, despite the limitations noted above. They also became aware of a nascent culture change in comparison to both Amicus and Unite members. It is the unequal access to power resources and the ability of these to make discrimination processes invisible that tends to undermine initiatives. Therefore, points of leverage would require the exposing of these processes, for example, how the 'rules of the game' were finessed to undermine the effectiveness of GPR and outflank women attempting to gain positional and personal power. Alliances with such as the stereotypical former NGA official identified in Chap. 7 would be invaluable in breaking with embedded masculine cultures. But of particular value would be exposing the almost entirely invisible politics of masculine identity, which is the mainstay of men's ability to control meanings. This is particularly important in view of the attempts to delegitimise gender initiatives on the basis that they are just 'gender identity politics' and inimical to class interests. The account given above should have a salutary effect and suggests that greater attention to such issues and the building of cross-gender solidarity would be a more fruitful direction for unions. Taking women activists' views (from the later research) on the need to make use of **all** resources to build stronger unions—that is, not privileging either lay members or officers over each other—would be a progressive step. But clearly a crucial element in all this would be the ending of the neoliberal era, which has proven so damaging to all workers, largely through the weakening of trade unions.

The potential to adopt any, or all, of these suggestions, plus many more that the author has no doubt overlooked, means that all is not lost—either for women or workers as a whole. Then there may come a time when women in print will no longer be 'paid Class 3, treated class [and so constrained to] act Class 3'.

References

Acker, J. (1998). The Future of 'Gender and Organizations': Connections and Boundaries. *Gender, Work and Organization, 5*(4), 195–206.

Acker, J. (2006). Inequality Regimes: Gender, Class, and Race in Organizations. *Gender and Society, 20*(4), 441–464.

BBC. (2013). Birmingham City Council to Settle 11,000 Equal Pay Claims. Retrieved November 30, 2017, from www.bbc.co.uk/news/uk-england-birmingham-24383352.

Bradley, H. (1996). *Fractured Identities: Changing Patterns of Inequality*. Cambridge: Polity Press.

Bradley, H. (1999). *Gender and Power in the Workplace*. Basingstoke: Macmillan Press.

Bundock, C. J. (1958). *The National Union of Printing, Bookbinding and Paper Workers*. London: Oxford University Press.

Child, J. (1967). *Industrial Relations in the British Printing Industry*. London: George Allen & Unwin.

Cockburn, C. (1983). *Brothers: Male Dominance and Technological Change*. London: Pluto Press.

Colgan, F., & Ledwith, S. (1996). Sisters Organising – Women and Their Trade Unions. In S. Ledwith & F. Colgan (Eds.), *Women in Organisations*. Basingstoke: Macmillan.

Colgan, F., & Ledwith, S. (2002). Gender, Diversity and Mobilisation in UK Trade Unions. In F. Colgan & S. Ledwith (Eds.), *Gender, Diversity and Trade Unions: International Perspectives*. London: Routledge.

Collinson, D., Knights, D., & Collinson, M. (1990). *Managing to Discriminate*. London: Routledge.

Connell, R. W. (1987). *Gender and Power*. Stanford, CA: Stanford University Press.

Crenshaw, K. (1991). Mapping the Margins: Intersectionality, Identity Politics and Violence Against Women of Color. *Stanford Law Review, 43*, 12–41.

Cunnison, S., & Stageman, J. (1995). *Feminizing the Unions*. Aldershot: Avebury.

Daniels, G., & McIlroy, J. (2009). *Trade Unions in a Neoliberal World: British Trade Unions under New Labour*. London: Routledge.

Dean, B. (2007). *Hot Mettle: SOGAT, Murdoch and Me*. London: Politico's Publishing.

Delafons, A. (1965). *The Structure of the Printing Industry*. London: MacDonald.

Dickens, L. (1997). Gender, Race and Employment Equality in Britain: Inadequate Strategies and the Role of Industrial Relations Actors. *Industrial Relations Journal, 28*(4), 282–291.

Drake, B. (1920). *Women in Trade Unions*. London: Virago Press.

Duffy, P. (2000). *The Skilled Compositor, 1850–1914: An Aristocrat Among Working Men*. Aldershot: Ashgate Publishing.

Edwards, P. K. (1995). The Employment Relationship and the Field of Industrial Relations. In P. K. Edwards (Ed.), *Industrial Relations: Theory and Practice*. Oxford: Blackwell Publishing.

Fosh, P., Morris, H., Martin, R., Smith, P., & Undy, R. (1993). Politics, Pragmatism and Ideology: The Wellsprings of Conservative Union Legislation 1979–1992. *Industrial Law Journal, 22*(1), 14–31.

Fraser, N. (2008). From Redistribution to Recognition: Dilemmas of Justice in a "Postsocialist" Age. In K. Olson (Ed.), *Adding Insult to Injury: Nancy Fraser Debates Her Critics*. London: Verso.

Gennard, J. (1990). *NGA: A History of the National Graphical Association*. London: Unwin Hyman.

Gennard, J., & Bain, P. (1995). *SOGAT: A History of the Society of Graphical and Allied Trades*. London: Routledge.

Gillespie, S. (1953). *A Hundred Years of Progress: The Record of the Scottish Typographical Association 1853 to 1952*. Glasgow: Robert Maclehose & Co.

The Guardian. (2010). The Most Hated Lawyer in Britain. Retrieved November 30, 2017, from www.theguardian.com/society/2010/may/01/stefan-cross-female-pay-birmingham.

Hacker, S. L. (1990). *Doing It the Hard Way*. London: Unwin Hyman.
Hartmann, H. (1981). The Unhappy Marriage of Marxism and Feminism: Towards a More Progressive Union. In L. Sargent (Ed.), *Women and Revolution: A Discussion of the Unhappy Marriage of Marxism and Feminism*. Boston: South End Press.
Howe, E. (1950). *The Federation of Master Printers 1900–1950*. London: Federation of Master Printers.
Hufton, O. (1995). *The Prospect Before Her: A History of Women in Western Europe, Volume 1 1500–1800*. London: HarperCollins.
Ledwith, S. (1991). *Sweet Talk, Soft Deals? – Gender and Workplace Industrial Relations*. Working Papers in Economics, Business and Management No. 35. London: Polytechnic of North London.
Lukes, S. (2005). *Power: A Radical View*. London: Palgrave Macmillan.
McDonald, J. R. (1904). *Women in the Printing Trades*. London: P. S. King and Son.
Munro, A. (1999). *Women, Work & Trade Unions*. London: Mansell.
Musson, A. E. (1954). *The Typographical Association*. London: Oxford University Press.
Noon, M. (2007). The Fatal Flaws of Diversity and the Business Case for Ethnic Minorities. *Work, Employment and Society, 21*(4), 773–784.
Phillips, A., & Taylor, B. (1980). Sex and Skill: Notes Towards a Feminist Economics. *Feminist Review, 6*, 79–88.
Reynolds, S. (1989). *Britannica's Typesetters: Women Compositors in Edinburgh*. Edinburgh: Edinburgh University Press.
Siltanen, J. (1994). *Locating Gender: Occupational Segregation, Wages, and Domestic Responsibility*. London: UCL Press.
Smith, P., & Morton, G. (2006). Nine Years of New Labour: Neoliberalism and Workers' Rights. *British Journal of Industrial Relations, 44*(3), 401–420.
Storey, J. (1983). *Managerial Prerogative and the Question of Control*. London: Routledge and Kegan Paul.
Thompson, P. (2011). The Trouble with HRM. *Human Resource Management Journal, 21*(4), 355–367.
Walby, S. (1986). *Patriarchy at Work*. London: Polity.
Walton, R., & McKersie, R. (1965). *A Behavioral Theory of Labor Negotiations*. New York: McGraw Hill.
Willis, P. (1977). *Learning to Labour*. Farnborough: Saxon House.
Zeitlin, J. (1985). Engineers and Compositors: A Comparison. In R. Harrison & J. Zeitlin (Eds.), *Divisions of Labour: Skilled Workers and Technological Change in Nineteenth Century Britain*. Brighton: Harvester Press.

Appendix: Characteristics of Survey Respondents and Interviewees

Table A.1a Characteristics of survey respondents

	Female	Male
Age		
0–10	1	1
10–20	3	3
20–30	65	38
30–40	148	146
40–50	177	218
50–60	177	166
Over 60	10	44
Non-response	14	5
Race		
White	569	597
Black African/Caribbean	3	3
African	6	1
Asian	6	6
Non-response/others[a]	11	14
Grade		
Class 1	95	373
Class 2	141	55

(*continued*)

T. Dawson, *Gender, Class and Power*, https://doi.org/10.1057/978-1-137-58594-3

Table A.1a (continued)

	Female	Male
Class 3	96	20
Different grading structure	132	118
Non-response	131	55
Job		
Finishing	314	146
Printing	24	239
Pre-press	36	95
Office	64	16
Others	139	112
Non-response	18	13
Employment status		
Full-time	501	611
Part-time	74	0
Casual	5	1
Temporary	2	2
Fixed-term contract	8	0
Others	0	0
Non-response	15	7
Company size		
Less than 5	15	17
6–19	61	82
20–99	213	243
100–499	256	249
Over 500	31	20
Non-response	19	10
Earnings status		
Sole earner	155	179
Main breadwinner	87	340
Neither	315	87
Non-response	38	15

Source: Survey of GPMU members 2002 (N = 1216 for each category)
[a]Responses that were either facetious or the small number of mixed race or alternative race respondents identified

Table A.1b Characteristics of survey respondents—geographical distribution

	Female	Male
Anglia	34	53
Birmingham and West Midlands	31	26
Central Midlands	58	39
Chiltern and Thames Valley	13	13
Cumbria	2	6
Greater London	12	21
Greater Manchester	16	21
Herts. and Essex	12	21
Home Counties West	7	13
Kent	12	21
Leeds and Central Yorkshire	56	31
Leicester and East Midlands	16	22
Merseyside, Central Lancs. and North Wales	20	30
Mid-Southern	19	16
North-east of England	19	19
North Lancs.	26	22
Norwich and North Anglia	15	23
Ouse Valley	12	28
Scotland	56	36
South Wales and Shires	24	26
South-West	35	56
South Yorkshire	13	8
Wessex	14	24
West Yorkshire	33	19
Wolds	21	11
Non-response	19	16

Source: Survey of GPMU members 2002 (N = 1216 for each category)

Table A.2 Characteristics of employers interviewed

Identifier	Sex	Company size	Location	Service in printing
Employer 1	Male	Federation	London/ countrywide	20+ years
Employer 2	Male	Small	W. Midlands	20+ years
Employer 3	Female	Multinational	Countrywide/ international	Less than a year
Employer 4	Male	Small	South-east	20+ years
Employer 5	Male	Small–medium	E. Midlands	20+ years

Table A.3 Characteristics of union officers interviewed

Identifier	National officer or branch officer	Male/female	Age range
Officer 1	National officer	Male	50–60
Officer 2	National officer	Male	40–50
Officer 3	National officer	Female	50–60
Officer 4	Branch officer	Female	40–50
Officer 5	Branch officer	Male	50–60
Officer 6	Branch officer	Male	50–60
Officer 7	Branch officer	Female	40–50
Officer 8	National officer	Female	60–65
Officer 9	National officer	Male	50–60
Officer 10	National officer	Male	50–60
Officer 11	Branch officer	Male	40–50
Officer 12	Branch officer	Male	50–60
Officer 13	National officer	Male	50–60
Officer 14	National officer	Male	60–65

Note: All the officers had service of at least ten years, many, more than 20 years. Further details would breach confidentiality

Table A.4 Criteria for selection of worker interviewees

Identifier	Sex	Age range	Family status	Branch	Job	Grade	Company size	Union membership (years)
Worker 1	Male	26	SE	Grtr. London	Finishing	Class 1	20–99	10
Worker 2	Female	23	N	Norwich and N. Anglia	Printing	Class 1	Less than 5	2
Worker 3	Male	46	MBW	North Lancs.	Printing	Class 1	100–499	28
Worker 4	Male	40	SE	Anglia	Finishing	Class 1	100–499	14
Worker 5	Male	33	N	Scotland	Labourer	N/A	100–499	13
Worker 6	Male	42	MBW	Leeds and Central Yorks.	Printing	N/A	100–499	24
Worker 7	Male	45	MBW	Central Midlands	Printing	Class 1	20–99	29
Worker 8	Male	45	SE	North Lancs.	Printing	Class 1	6–19	28
Worker 9	Male	35	MBW	Merseyside, Central Lancs. and N. Wales	Finishing	Class 1	100–499	15
Worker 10	Male	40	MBW	Herts. and Essex	Pre-Press	Class 1	20–99	20
Worker 11	Male	37	SE	Home Counties West	Finishing	Class 1	100–499	14
Worker 12	Female	38	N	NE England	Other	Class 2	100–499	21
Worker 13	Female	47	MBW	West Yorkshire	Finishing	Class 3	20–99	30
Worker 14	Female	31	–	Grtr. Manchester	Other	N/A	100–499	37
Worker 15	Female	38	SE	Leicester and E. Midlands	Finishing	Class 2	Over 500	2
Worker 16	Female	34	–	Merseyside, Central Lancs. and N. Wales	Finishing	Class 1	20–99	1
Worker 17	Female	42	N	Birmingham and W. Mids.	Pre-press	Class 1	6–19	1
Worker 18	Female	34	N	West Yorkshire	Other	Class 2	–	17
Worker 19	Female	35	N	Wessex	Finishing	Class 3	100–499	15
Worker 20	Female	31	MBW	Anglia	Pre-press	N/A	100–499	14

(continued)

Table A.4 (continued)

Identifier	Sex	Age range	Family status	Branch	Job	Grade	Company size	Union membership (years)
Worker 21	Male	60	SE	Leeds and Central Yorks.	Finishing	N/A	100–499	17
Worker 22	Male	52	MBW	Chiltern and Thames Valley	Printing	Class 1	20–99	36
Worker 23	Male	53	N	Central Midlands	Pre-press	Class 1	6–19	38
Worker 24	Male	51	SE	Home Counties West	Finishing	Class 1	20–99	35
Worker 25	Male	59	SE	Ouse Valley	Printing	Class 1	Over 500	31
Worker 26	Female	54	SE	Kent	Pre-press	Class 1	20–99	33
Worker 27	Female	63	MBW	Norwich and N. Anglia	Finishing	Class 3	20–99	N/A
Worker 28	Female	55	SE	Grtr. Manchester	Finishing	Class 2	20–99	40
Worker 29	Female	61	N	South Yorkshire	Finishing	Class 3	Over 500	15
Worker 30	Female	54	SE	Scotland	Pre-press	N/A	20–99	2

Key:

Family status: *SE* sole earner, *MBW* main breadwinner, *N* neither

Grade: Class 1, Class 2, Class 3, N/A (for those who have a different grading structure)

Notes:

Worker 12 is a machine operator; Worker 14 works in direct mailing; Worker 18 crosses boundaries, that is, finishing/administration/warehouse

Worker 27 had interrupted union membership

Table A.5 Characteristics of union officers and activists interviewed: 2014–2017

Identifier	Union position	Male/female	Age
R2(1)	Former union rep	Female	60–70
R2(2)	Former union rep	Female	80–90
R2(3)	Former union rep	Female	50–60
R2(4)[a]	Retired union officer	Female	60–70
R2(5)	Former union rep	Female	50–60
R2(6)[a]	Former union officer	Male	60–70
R2(7)	Union FTO	Female	50–60
R2(8)[a]	Union FTO	Female	50–60
R2(9)	Former union FTO	Female	50–60

[a]Also interviewed in 2002

Glossary

The division of labour in printing breaks down into three parts, and the terminology relating to that is explained below:

CTP that is, computer-to-plate technology refers to the computerisation of the pre-press process, allowing the printing plates to be generated direct from the computer process when this would have been a separate process previously—see planning and platemaking, below. One interviewee provided an example of how revolutionary this process has been when he explained that a brochure that used to be produced by his company would have taken about six weeks using old technology but now only takes a few days. There is also computer-to-print technology which is generally referred to as digital printing where the whole process is computer generated—there are no plates.

Finishing or bindery work all processes that are necessary to produce a finished product, for example, cutting and creasing, guillotining, collating, folding, sewing, gluing, packing and so on. The work that women did in the bindery was often handwork traditionally, and this is sometimes referred to as table work. Women may be referred to as bindery or finishing assistants.

Hot metal composition comprised foundry workers casting type, that is, letters and so on, that would then be made up into forms that may cover, perhaps, 16/24/48 pages each.

Letterpress printing from a raised relief surface using metal types and half-tone blocks (images).

T. Dawson, *Gender, Class and Power*, https://doi.org/10.1057/978-1-137-58594-3

Lithography (litho) planographic printing from a flat surface reliant on the mutual antipathy of greasy ink and water. The printing plate is successively inked and damped to prevent portions of the surface accepting ink.

Photocomposition replaced 'hot metal' composition and comprised production of film by photographic methods for different printing methods, for example, letterpress, litho or gravure.

Planning/platemaking part of the pre-press process—comprising assembly of individual pages of photographic film to suit particular presses, that is, 8/16/32 pages on each plate.

Pre-press/origination all work preparatory to the printing process such as typesetting, page lay-out and platemaking—the aim is to transfer words, photographs, graphics and so on onto a printing surface.

Printing that is, all processes necessary for the actual printing of typeset material. Two types of worker operated in this department—the machine minders, that is, craftsmen, and their machine assistants, that is, semi-skilled male workers. Women were only present where they were required to take material off machines.

Rotaprint small litho machine used originally for office reproduction. Initially the unions wanted to keep women off these machines, but later, when they started appearing in offices, they tried unsuccessfully to recruit the women who were often already in public sector or white-collar unions.

Typesetting formatting and/or input of words and images to a computer.

Index[1]

[1] Note: Page number followed by 'n' refer to notes.

T. Dawson, *Gender, Class and Power*, https://doi.org/10.1057/978-1-137-58594-3

Gender studies

Lightning Source UK Ltd.
Milton Keynes UK
UKHW02n1037030818
326716UK00009B/407/P

9 781137 585936